Quantitative Neurophysiology
Revised Edition

Quantitative Neurophysiology, Revised Edition
Joseph V. Tranquillo

ISBN: 978-3-031-00500-8   paperback
ISBN: 978-3-031-01628-8   ebook

DOI 10.1007/978-3-031-01628-8

A Publication in the Springer series
SYNTHESIS LECTURES ON BIOMEDICAL ENGINEERING
Lecture #21
Series Editor: John D. Enderle, University of Connecticut

Series ISSN
Synthesis Lectures on Biomedical Engineering
Print 1930-0328   Electronic 1930-0336

Synthesis Lectures on Biomedical Engineering

Editor
John D. Enderle, *University of Connecticut*

Artificial Organs
Gerald E. Miller
2006

Signal Processing of Random Physiological Signals
Charles S. Lessard
2006

Image and Signal Processing for Networked E-Health Applications
Ilias G. Maglogiannis, Kostas Karpouzis, and Manolis Wallace
2006

Quantitative Neurophysiology

Revised Edition

Joseph V. Tranquillo
Bucknell University

SYNTHESIS LECTURES ON BIOMEDICAL ENGINEERING #21

ABSTRACT

Quantitative Neurophysiology is supplementary text for a junior or senior level course in neuro-engineering. It may also serve as an quick-start for graduate students in engineering, physics or neuroscience as well as for faculty interested in becoming familiar with the basics of quantitative neuroscience. The first chapter is a review of the structure of the neuron and anatomy of the brain. Chapters 2-6 derive the theory of active and passive membranes, electrical propagation in axons and dendrites and the dynamics of the synapse. Chapter 7 is an introduction to modeling networks of neurons and artificial neural networks. Chapter 8 and 9 address the recording and decoding of extracellular potentials. The final chapter has descriptions of a number of more advanced or new topics in neuroengineering. Throughout the text, vocabulary is introduced which will enable students to read more advanced literature and communicate with other scientists and engineers working in the neurosciences. Numerical methods are outlined so students with programming knowledge can implement the models presented in the text. Analogies are used to clarify topics and reinforce key concepts. Finally, homework and simulation problems are available at the end of each chapter.

KEYWORDS

Neurophysiology, neuroengineering, neuroscience, electrophysiology, biomedical engineering, neuron, neural signal processing, neural anatomy, brain, CNS

Contents

Preface

The function of the brain and connection to the mind is perhaps one of the most enigmatic pursuits in all of science. Some have even dubbed the study of the brain the final frontier. Others claim that we will never know exactly how our brain works. We do, however, understand a great deal about the brain, as evidenced by nearly 100 years of Nobel Prizes

Year	Winners	For
1906	Golgi and Cajal	Structure of the nervous system
1920	Nernst	Membrane potentials
1932	Sherrington and Adrian	The synapse
1936	Dale and Loewi	Chemical propagation across a synapse
1944	Erlanger and Gasser	Differentiation of nerve fibers
1949	Hess	Role of interbrain in regulation
1950	Kendall, Reichstein and Hench	Hormones in the adrenal cortex
1963	Eccles, Hodgkin and Huxley	Ionic mechanisms of action potentials
1970	Katz, Euler and Axelrod	Function of neurotransmitters
1977	Guillemin and Schally	Neurohormones in the brain
1981	Sperry, Hubel and Wiesel	Specialization of cerebral hemispheres
1991	Neher and Sakmann	Function of single ion channels
1994	Gilman and Rodbell	G-proteins in signal transduction
1997	Prusiner	Prions as a principle for infection
2000	Carlsson, Greengard and Kandel	Dopamine signal transduction
2003	Lauterbur and Mansfield	Magnetic Resonance Imaging (MRI)
2003	Agre and MacKinnon	Ion channel selectivity

The objective of "Quantitative Neurophysiology" is to provide an overview of the theory underlying electrical impulses in the neuron. The material presented is purposely distinct from a traditional neuroscience text in that it is geared toward engineers and applied scientists. As such the focus will be place more on neurons in the central nervous system, where a deep quantitative understanding has been developed, and less on the large-scale functions of the brain. Despite this narrow focus you will find some classic neuroscience woven into the text.

For the student, it is hoped that you will see concrete applications of fundamental engineering and science concepts in the description of a real biological system. It will be most helpful if you have already taken differential equations as well as a basic circuits course. The philosophy of the text is that by understanding the mathematical models of neuronal function (Chapters 2-6), you will gain a deeper understanding of how neurons work together to create complex behavior (Chapter 7) and how we might listen in on the chatter of neurons to unravel their language (Chapters 8-9). Finally, the theory presented should prepare you to read the latest neuroengineering literature and move on to more advanced topics (Chapter 10).

For the instructor, it is hoped that this text will provide a resource that can aid you in your instruction. I have found that students are capable of digesting one chapter each week ($\approx$3 hours of class time, supplemented by homework problems). At the end of each chapter are homework and simulation problems which can be assigned as is, or modified to better suit the abilities of the students. This schedule leaves ample time to cover more advanced topics at the instructor's discretion.

This text would not be possible without the support of two important groups. First, I wish to thank the Bucknell students who have taken my Neural Signals and System course. Second, I wish to thank my family for their patience and keeping me focused on what is really important in life.

Joseph V. Tranquillo
Lewisburg, Pennsylvania

CHAPTER 1

Neural Anatomy

The brain may be studied at many scales, ranging from molecular to psychological. In this text, the focus will be primarily on quantifying individual cells and groups of cells for two reasons. First, much more is known about the function of individual neurons. Second, an underlying assumption of neuroscience is that a complete understanding of brain function will require deep insights at the cell and molecular level. In this chapter, the anatomy of the neuron is explored to the depth needed to understand future chapters. It is very important, however, to not lose sight of the goal of understanding the brain at the larger scales. Therefore, this chapter also outlines some of the gross level structures of the brain.

1.1   THE NEURON

The human brain is composed of more than 10^{12} neurons. Table 1.1 below is a summary of some types of neurons.

Table 1.1:	
Bipolar	Found in eye, transduction of light to neural impulse
Somatosensory	Found in skin, tactile senses of pain and proprioception
Motor	Found in spinal cord, projects to muscles
Pyramidal cell	Found in cortex, relay information within brain
Purkinje Cell	Found in cerebellum, motor skills
Association cell	Found in thalamus, connect neurons together

Although the anatomy and function of neurons vary throughout out the brain, Fig. 1.1 shows the features of a generic neuron. Like other cells, the neuron separates space inside from space outside by a $5nm$ bilipid membrane that acts as an electrical insulator. Typically, the neuron is divided into the dendritic tree, soma and axon.

The *dendritic tree* is a complex web of branching structures that range in diameter from $1-20\mu m$. One interesting feature of some dendrites is the presence of *spines* that play a somewhat unclear electrical or chemical role in neuronal function.

The *soma*, or cell body, is approximately $20\mu m$ in diameter and contains most of the organelles, including the nucleus, Golgi apparatus, mitochondria, microtubuals, and endoplasmic reticulum. It is here that the cell generates ATP, packages neurotransmitters, houses the genetic material, and assembles proteins for the cell.

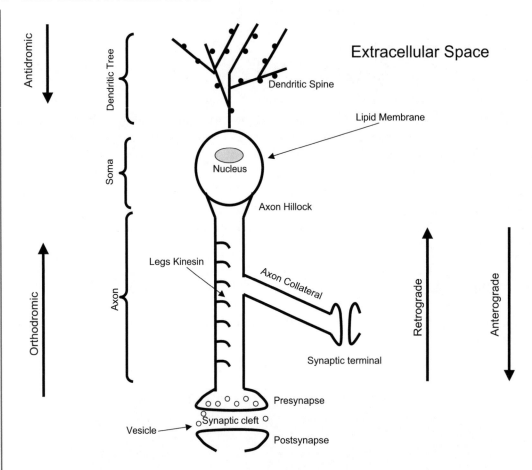

Figure 1.1: Schematic of a typical neuron.

The *axon* branches off of the soma at the *axon hillock*. Axons can vary greatly in length ($1mm -$ $1m$) and diameter ($1 - 25\mu m$). The intracellular space of the axons are covered in small specialized proteins called *legs kinesin* which mechanically transport ATP, vesicles filled with neurotransmitter and enzymes to and from the soma. Some axons have offshoots called *axon collaterals*.

The *synapse* is the site where the dendrite of a neuron interfaces with the axon of another neuron. It was first described in 1897 by Sherrington who coined the term from the Greek, Syn (together) + Haptin (to fasten). There are approximately 10^{15} synapses in the brain, so each neuron has on average 1,000 synapses. The number of synapses in an individual neuron, however, can vary greatly. It is important to note that the synapse is not a singular structure, but rather the combination of three structures. The *pre-synapse* is the very end of an axon and houses *vesicles*, or small spheres of membrane, that contain *neurotransmitters*. The *post-synapse* is on the very end of a dendrite. The

20nm space between the pre-synapse and post-synapse is called the *synaptic cleft* and is technically outside of both neurons.

Neurotransmitters are specialized molecules that are packaged into vesicles in the soma, transported to the end of the pre-synaptic axon by the legs kinesin, and released into the synaptic cleft in response to an electric impulse. Neurotransmitters diffuse across the synaptic cleft and reach the post-synapse where they either excite or inhibit electrical impulses in the dendrite of a new neuron. There is enormous diversity in the molecules that function as neurotransmitters as well as the mechanisms by which they affect the post-synapse.

To indicate the direction of ionic or molecular movement in a neuron, two sets of terms have been defined. When discussing motion within the axon, *anterograde* is movement from the *anterograde* soma to the end of the axons and *retrograde* is movement from the axon to the soma. To discuss movement in the entire neuron, *orthodromic* is in the direction from the axon to the dendrites, while *antidromic* is from the dendrites to the axon.

1.2   GLIAL CELLS

Although often overlooked, *glial cells* in the brain outnumber neurons nearly two to one. Historically, glial cells were thought to only impact electrical properties indirectly by maintaining extracellular ion concentrations and speeding electrical propagation through the growth of the *myelin sheath*. Recent evidence, however, suggests that glial cells may play a more direct role in the propagation of impulses. Some of the more common types of glial cells are listed in Table 1.2.

Table 1.2:	
Astroglia	Structural support and repairs
Oligodendroglia	Insulate and speed transmission in CNS
Schwann Cells	Insulate and speed transmission in PNS
Microglia	Perform phagocytosis of damaged or dead cells
Ependymal cells	Line the ventricles and produce cerebrospinal fluid

1.3   THE BRAIN

While our primary focus will be on quantifying the function of nerve cells, it is important not to lose sight of the fact that our brain regulates nearly all voluntary and involuntary actions in our bodies. It is often reported that we only use 10% of our brain, with the implication that there is 90% of the brain that we do not understand. The 10% number is the percentage of the brain that is active *at any one time*. Each structure of the brain is known to perform some function. What is less clear is how these structures relate to one another and how they use neurons to perform their functions. Below is a brief overview of the structure and function of the most significant regions of the brain.

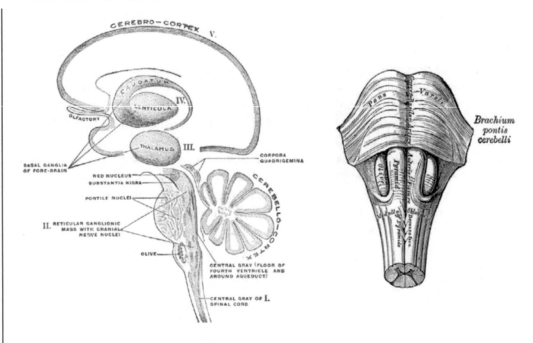

Figure 1.2: Gross anatomy of the brain.

1.3.1   From Neuron to Nuclei

Typically, cells that are close to one another are more densely connected by synapses and perform similar functions. These groupings of cells form a *nucleus*. In histological slices, the dense tangle of cell bodies, dendrites, and capillaries have a grayish color and so have been given the name *grey matter*. Several nuclei may be connected together to form larger structures in the brain. The connections between nuclei are made by bundles of long axons, called *tracts*, that may carry large amounts of information. Long axons require fast propagation and so are covered in a fatty coating of myelin. In histological slices, these dense regions of axons have a white color and so have been given the name *white matter*.

1.3.2   Brain Systems

The gross anatomy of the brain consists of both tissue (neurons and glia) as well as chambers called ventricles that are filled with cerebral-spinal fluid. The tissue is organized into three regions based upon embryonic development.

The *hindbrain*, or Rhombencephalon (Table 1.3), is an offshoot of the spinal cord and was inherited from the reptile brain. It is primarily involved in involuntary control of basic functions such as breathing and heart rate. In addition to forming the lower portion of the brain stem, it also includes the cerebellum, a relatively large outcropping at the base of the brain that controls motor movements and relays signals to the spinal cord.

Table 1.3:		
Myelencephalon		Medulla oblongata
Metencephalon		Contains 4th Ventricle
	Pons	Relay sensory information
	Cerebellum	Integration of sensory and motor
	Reticular Formation	Rudimentary eye movements

The *midbrain*, or Mesencephalon, forms the upper part of the brain stem. Its primary function is to connect the lower brain stem with higher-level brain structures. In size, it is much smaller than the other two embryonic regions. It is involved in involuntary motor control and sensation.

The *forebrain*, or Prosencephalon (Table 1.4), forms the bulk of the brain tissue in mammals. It contains the limbic system that regulates drives, hunger, hormones, emotions, and memory, as well as the cerebral cortex. The regions of the forebrain are tightly connected together by tracts in complex feedback networks.

Table 1.4:		
Diencephalon		Lateral and 3rd ventricles, limbic system
	Epithalamus	Pineal body, wake/sleep patterns
	Thalamus	Relay information to cerebral cortex
	Hypothalamus	Link to endocrine system, metabolism, hunger
	Pituitary Gland	Master endocrine gland
Telencephalon		Also called the cerebrum
	White Matter	Axons that connect cortex to other structures
	Amygdala	Process memory and emotions
	Hippocampus	Process memory and spatial navigation
	Rhinencephalon	Olfaction (smell)
	Cerebral Cortex	Higher level tasks

The *cerebral cortex* (Table 1.5) occupies the top $2-4mm$ of the forebrain and is composed of grey matter organized into six layers. The cortex is split into a number of *lobes* which are roughly mirrored on left and right *hemispheres* of the brain.

1.3.3 Blood Brain Barrier

In the body, capillaries are lined with endothelial cells that contain small gaps, allowing for easy diffusion of chemicals and gasses to and from the blood. In the brain, a similar membrane, called the *blood brain barrier*, is present but the cells are packed more tightly so that most chemicals can not

Structure	Location	Notes
Frontal Lobe	Behind forehead	Impulse control, language
Parietal Lobe	Crown of head	Sensory, numbers
Temporal Lobe	Behind temples	Auditory, speech, and vision
Occipital Lobe	Back of head	Primary visual center
Insula	Between temporal and parietal	Emotions, pain, addition
Cingulate Cortex	Surrounds the corpus collosum	Emotions
Corpus Collosum	Connection between hemispheres	Millions of axons

Table 1.5:

diffuse through the membrane. The only chemicals that can pass are either lipid soluble molecules (e.g., O_2, CO_2, ethanol, steroid hormones) or those carried by a specific channel (e.g., sugars, amino acids). The result is that the brain is able to maintain a high rate of metabolism but is protected from potentially dangerous chemicals and infections. In addition, glial cells surround capillaries in the brain, forming a second barrier. There are, however, a few entry points that allow the brain to sense the concentrations of hormones in the blood. The downside of the blood brain barrier is that it is difficult to design drugs that cross the barriers and work directly on the brain.

1.3.4   Directions in the Brain
Directions in the brain are the same as in the body. Anterior refers to the front of the brain while posterior is toward the back. Medial is toward the center of the brain while lateral is away from the center. Superior (also dorsal) is toward the top of the brain while interior (also ventral) is toward the base of the brain.

Two additional terms are used in the brain that trace a curved line from the bottom of the spinal cord to the nose (or frontal lobe). Moving in a *rostral* direction is moving toward the nose. Moving in a *caudal* direction is moving toward the bottom of the spinal cord.

1.3.5   Inputs and Outputs
The focus of the text is on neurons in the central nervous system. The mathematical models, however, apply equally well to neurons in the peripheral nervous system. We therefore briefly mention that the brain interacts with neurons in the body. Neurons that carry impulses to the brain are called *afferent*, sensory or receptor nerves. Typically, these nerves originate in a sensory system such as the eyes or nose. Neurons that carry impulses away from the brain are called *efferent*, motor or effector nerves. Typically, they will innervate muscles or glands.

1.3.6   Dynamic Brain Systems
It should be clear from the brief overview above that the brain is a very complex and hierarchical system. The most complex aspect of the brain, however, is that it is constantly changing. First,

during development new neurons are rapidly being formed. There is, however, some evidence to suggest that new neurons form at a slower rate even in older brains. Second, two neurons may form a new connection via a synapse or a gap junction. Third, the strength of existing connections may be changed by an increase in synaptic receptors or number of gap junctions. Fourth, two neurons may cut their connection if it is no longer needed. Fifth, neurons are constantly dying. Sixth, glial cells may connect to one another and to the network of neurons. Lastly, the vascular blood supply must reach within $100\mu m$ of every cell in the brain. Your image of the brain should therefore be one of three interacting networks (neurons, glial and vasculature) all of which are constantly changing. It is incredible that the brain can maintain its basic functions in the face of this change.

CHAPTER 2

Passive Membranes

Electrophysiology is the study of the electrical properties of biological materials, from molecules to the entire body. Although all materials in the body can be characterized by their electrical properties, the nervous and muscular system in particular use electrical impulses to communicate information between cells. In this chapter, we will explore the basic principles of cellular electrophysiology which may be applied to any cell in the body. In Ch. 3, we will consider the more specialized electrical properties of neurons.

2.1   CELLULAR ELECTROPHYSIOLOGY

As we are primarily concerned with electrical properties, we must first define what voltages (difference in potential) and currents (flow of charged particles) mean in the context of a cell. All of the normal principles of electricity apply, however, most electrical texts consider the movement of negatively charged electrons. In biological systems, currents are a flow of ions (e.g., Na^+, Cl^-) and voltages are differences in potential created by gradients in ionic concentrations.

2.1.1   Cellular Voltages
The *transmembrane voltage*, V_m, is defined as the difference in potential across the cell membrane (Fig. 2.1) and is typically measured in mV.

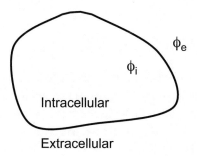

$$\phi_e$$
$$\phi_i$$
Intracellular

Extracellular

Figure 2.1: Definition of membrane voltage, V_m.

$$V_m = \phi_i - \phi_e = -\Delta\phi \tag{2.1}$$

where ϕ_i is the potential inside the cell and ϕ_e is the potential outside the cell. Surprisingly, the transmembrane voltage will reach a steady state called the *resting membrane voltage*, V_m^{rest}, that is not zero. In most neurons $V_m^{\text{rest}} \approx -60mV$, meaning that at rest the inside of the cell has a more negative potential than the outside. Any positive change in V_m from rest is called a *depolarization* (Fig. 2.2) and may occur because the inside of the cell has become more positive or the outside of the cell has become more negative. Any negative change in V_m from rest is called a *repolarization*. Membrane voltages below rest, are referred to as *hyperpolarized*.

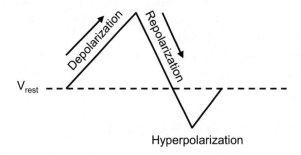

Figure 2.2: Depolarization, repolarization, and hyperpolarization of V_m.

2.1.2   Cellular Currents

The *transmembrane current*, I_m, is a measure of the movement of ions across the membrane and is typically measured in units of $\mu A/cm^2$. By definition, a positive I_m is positive charge leaving the cell. Therefore, a positive I_m will repolarize the cell membrane voltage. The transmembrane current is the sum of four types of currents

$$I_m = I_{cm} + I_{\text{ion}} + I_{\text{syn}} - I_{\text{stim}} . \tag{2.2}$$

The *capacitive current*, I_{cm}, is a result of the natural capacitance of the cell membrane. Recall that a capacitor is nothing more than an insulator sandwiched between two conductors. The cell membrane is composed of lipids which are a natural isolator. The intra and extracellular solutions are salty water and therefore are good conductors. We will examine the role of the membrane capacitance, C_m, in more detail in Sec. 2.1.4.

The *ionic current*, I_{ion}, is the primary way for ions to cross the membrane. I_{ion} may be composed of many different currents (carrying different ions) summed together. We will consider I_{ion} in greater detail in this chapter and Ch. 3.

The *synaptic current*, I_{syn}, is any one of hundreds of specialized currents that allow neurons to communicate chemically across a synapse. We will consider I_{syn} in Ch. 6.

The *stimulus current*, I_{stim}, is an externally applied current and will be considered in more detail in Sec. 2.2. Note that the sign of I_{stim} is opposite of the other currents.

2.1.3 Membrane Circuit Analog

It is customary to use circuit analogies to describe the relationship between voltages and currents. In Fig. 2.3, the top node represents the inside of the cell and the bottom node represents the outside of the cell. All currents are represented by a unique parallel pathway through which current may cross the membrane. In the literature, this circuit is therefore known as the *parallel conductance model*. By conservation of current, the total membrane current, I_m, as computed at the top node, must sum to zero.

$$0 = I_{cm} + I_{ion} + I_{syn} - I_{stim} \;. \tag{2.3}$$

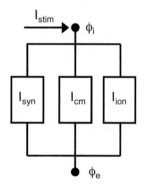

Figure 2.3: Parallel conductance model.

Substituting the known relationship between voltage and current for a capacitor,

$$0 = C_m \frac{dV_m}{dt} + I_{ion} + I_{syn} - I_{stim} \tag{2.4}$$

$$\frac{dV_m}{dt} = \frac{1}{C_m} \left[-I_{ion} - I_{syn} + I_{stim} \right] \;. \tag{2.5}$$

Equation (2.5) is a differential equation that describes how V_m evolves over time based upon the currents that flow across the cell membrane.

2.1.4 The Membrane Capacitance

The value of the membrane capacitance can be directly computed by

$$C_m = \frac{k\epsilon_0}{d} \tag{2.6}$$

where k is the dielectric constant of the insulator, ϵ_0 is the permittivity of free space($1 \times 10^{-9}/36\pi \ F/m^2$), and d is the membrane thickness. Using $k = 3$ (value for oil) and $d = 3nm$:

$$C_m = \frac{3 \times 10^{-9}}{36\pi(3 \times 10^{-9})} = 0.009\frac{F}{m^2} = 0.9\frac{\mu F}{cm^2} \ . \tag{2.7}$$

For simplicity, $1\mu F/cm^2$ is often used as an approximation for C_m. As our estimate for C_m is measured per unit area, it is independent of the size of the cell and therefore a biophysical property of any cell surrounded by a bilipid layer.

2.2   STIMULATING THE PASSIVE MEMBRANE

If an external current (I_{stim}) is applied to the membrane of a cell, as in Fig. 2.4, such that positive ions are forced into the cell, the membrane voltage will depolarize. For now we will assume that there are no synapses ($I_{\text{syn}} = 0$), so

$$\frac{dV_m}{dt} = \frac{1}{C_m}[-I_{\text{ion}} + I_{\text{stim}}] \ . \tag{2.8}$$

The membrane, although composed primarily of lipids, has "leaky" channels that will allow some current to pass. Experimentally, it has been observed that when V_m is close to V_m^{rest}, the leakage of ions is proportional to V_m. Therefore, we can approximate I_{ion} using Ohm's Law:

$$I_{\text{ion}} = \frac{\phi_i - \phi_e}{R_m} = \frac{V_m}{R_m} \tag{2.9}$$

where R_m is the *specific membrane resistivity* ($k\Omega cm^2$) to current flow and measures the "leakiness" of the membrane. Given our assumption of linear R_m,

$$\frac{dV_m}{dt} = \frac{1}{C_m}\left[-\frac{V_m}{R_m} + I_{\text{stim}}\right] \ . \tag{2.10}$$

Rearranging Eq. (2.10)

$$R_m C_m \frac{dV_m}{dt} + V_m = R_m I_{\text{stim}} \ . \tag{2.11}$$

It has become customary rewrite Eq. (2.11) as:

$$\tau_m \frac{dV_m}{dt} + V_m = V_\infty \tag{2.12}$$

where

$$\tau_m = R_m C_m \tag{2.13}$$

is the subthreshold *membrane time constant* measured in *msec* and

$$V_\infty = R_m I_{\text{stim}} \tag{2.14}$$

is the steady-state voltage as time $\rightarrow \infty$. It is important to note that V_∞ is always measured relative to V_m^{rest}.

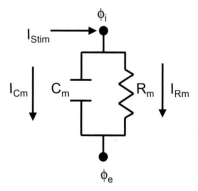

Figure 2.4: RC circuit analog of a passive cell membrane.

When I_{stim} is applied, V_m will charge up the capacitor to V_∞ at a rate governed by τ_m. If the current remains on, the voltage will remain at V_∞. Solving Eq. (2.12) during stimulation,

$$V_m(t) = V_\infty(1 - e^{-t/\tau_m}) \tag{2.15}$$

where t is the time after applying I_{stim}.

When I_{stim} is turned off, V_m will have charged to some initial voltage V_0. From this initial value, V_m will return to V_m^{rest}, again at a rate governed by τ_m

$$V_m(t) = V_0 e^{-t/\tau_m} . \tag{2.16}$$

2.2.1   Finding Membrane Properties from $V_m(t)$

The left-hand side of Fig. 2.5 is a plot of $V_m(t)$ over time as a current pulse is applied and then removed. The rising phase is governed by Eq. (2.15) while the falling phase is governed by Eq. (2.16). The rising or falling phase can be used to determine the membrane properties, R_m, C_m, and τ_m. The right-hand side of Fig. 2.5 shows a stimulus that is applied only for a short time, so V_m does not have time to reach V_∞ before the stimulus is turned off.

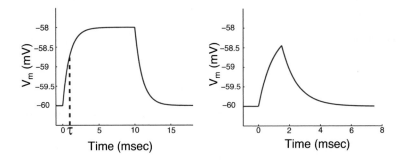

Figure 2.5: Charging and discharging of a passive membrane.

2.2.2   The Passive Membrane

To perform the analysis above, we assumed that the current flowing through the membrane was linearly proportional to the membrane voltage. It turns out that this assumption is a good approximation as long as V_m is below some *threshold* voltage, V_m^{th}. When $V_m < V_m^{th}$, the membrane is called *passive*, R_m is a constant and the membrane can be represented as the RC circuit in Fig. 2.4. In a real neuron, V_m^{th} is typically between 5mV and 10mV higher than V_m^{rest}. When V_m depolarizes above the threshold, however, the membrane will become *active* and R_m will no longer be a constant. In Ch. 3 we will consider the nonlinear relationship between R_m and V_m above the threshold.

2.3   STRENGTH-DURATION RELATIONSHIP

To perform our analysis of the passive membrane we assumed that $V_m < V_m^{th}$. It would therefore be helpful to know in what situations our assumption is valid. Examining Eq. (2.15), $V_m(t)$ during the stimulus is dependent on $V_\infty = I_{\text{stim}} R_m$ and $\tau_m = R_m C_m$. Consider that I_{stim} may be small such that $V_\infty < V_m^{th}$. In this case, the stimulus may remain on forever and V_m will charge up to a steady state value below V_{th}. In other words, the membrane will remain passive. If I_{stim} is steadily increased, eventually V_∞ will be equal to V_m^{th}. In this case, the steady state will level out exactly at the threshold but it may take a very long time for V_m to reach threshold. If I_{stim} is increased further, V_∞ will clearly be greater than V_m^{th}. As the right side of Fig. 2.5 indicates, however, even if $V_\infty > V_m^{th}$ the stimulus may be on for such a short time that the membrane does not have time to charge to the threshold. It follows that there is an interplay between the strength of I_{stim} and the duration of I_{stim}, which can be represented graphically in a *strength-duration* plot.

Given a constant value for V_m^{th}, Fig. 2.6 shows the combinations of strength and duration that exactly charge the membrane to the threshold. Mathematically, we can set $V_m = V_m^{th}$ and substitute into Eq. (2.15)

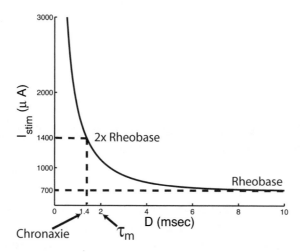

Figure 2.6: Strength-duration curve.

$$V_m^{th} = I_{\text{stim}} R_m (1 - e^{-D/\tau_m}) \tag{2.17}$$

where D is the duration of the stimulus. One interesting limit is to find the minimum current that can bring the membrane to V_m^{th}. If $D \to \infty$

$$V_m^{th} = I_{\text{stim}} R_m \tag{2.18}$$

$$I_{\text{stim}} = I_{\text{rhe}} = \frac{V_m^{th}}{R_m} \tag{2.19}$$

which explains the value of the asymptote in Fig. 2.6. This value is called *rheobase*, I_{rhe}, and is a measure of current. Another important measure can be derived by assuming that we apply two times the rheobase current to find the corresponding time to charge the membrane to V_m^{th}.

$$V_m^{th} = 2I_{\text{rhe}} R_m (1 - e^{-D/\tau_m}) \tag{2.20}$$

and solving for D

$$D = -\tau_m \cdot ln\left[1 - \frac{V_m^{th}}{2I_{\text{rhe}}R_m}\right] \tag{2.21}$$

but we know that $I_{\text{rhe}}R_m = V_m^{th}$ so

$$D = -\tau_m \cdot ln\left[1 - \frac{1}{2}\right] \tag{2.22}$$

$$T_c = D = 0.693\tau_m . \tag{2.23}$$

The time T_c is known as *chronaxie*. Therefore, given a strength duration curve, and Eqs. (2.19) and (2.23) one could find the passive membrane properties.

2.3.1   A Fluid Analogy

An alternative way to think about the passive membrane is as a cup with a hole in the bottom and a line painted half way up the side. In the analogy, the volume of the cup is similar to the capacitance in that it can store the quantity of interest (water instead of charge). The size of the hole is similar to the resistance in that it tends to leak out the quantity of interest (water instead of charge). The line on the cup represents the threshold voltage. As long as the water level remains below the threshold, the water will leak from the hole. A stimulus (a flow rate of charges) in our analogy would be the steady poring of water into the cup.

Using our analogy we can gain some intuition about charging and discharging in response to a stimulus. For example, if water is poured in slower than the rate at which it leaks out of the hole, then the threshold line will never be reached. This is analogous to being below I_{rhe}. You could imagine, however, that more aggressive pouring would eventually result in the threshold being reached but only very slowly. It may also be possible to pour water in only for some short period of time. If this is the case, then the rate at which you pour and the duration of your pour will determine if the water will reach the threshold. You could in fact create a strength-duration plot for this cup-water system. Furthermore, if you assume that the cup already has water in it, you may ask how long it will take for the cup to drain. The two parameters of interest will be the size of the cup (capacity) and the size of the hole (resistance). The combination of the two, as in the circuit analogy, can be used to define the time constant of the filling or draining of the cup.

2.4   THE MEMBRANE AT REST

A careful examination of Fig. 2.4 will reveal that V_m^{rest} must be $0mV$, which biologically is not true. In this section we will examine how a cell can maintain a nonzero resting potential.

2.4.1 The Forces Driving Ion Movement

Consider Fig. 2.7 where circles and dots represent two types of positively charged ions but the membrane will only allow dots to pass through. The special property of the membrane to easily pass some ions but not others is called *selective permeability* and results in a nonzero resting potential.

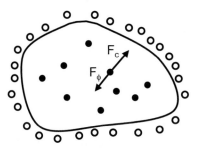

Figure 2.7: Concentrations and electrical potentials in a cell.

There are two forces driving the motion of an ion. The electric force, F_ϕ, is due to any difference in potential ($\Delta\phi$) between the inside and outside of the cell. Remember that potential differences arise because there are different amounts of charge on either side of the membrane. The potential is therefore due to *all* of the charges, whether they are able to cross the membrane or not. The chemical force, F_C, is due to differences in *specific* ionic concentrations (ΔC) across the membrane and will act only on a single type of ion (e.g., dot or circle in Fig. 2.7). It is because F_ϕ is a function of all charges and F_C is a function of only a specific ion that the membrane can support a nonzero rest potential.

2.4.2 A Helium Balloon Analogy

In a cell, there are two forces pulling on charged particles that balance out. The same concept can be applied to the balance of two forces acting on a Helium balloon. There are always two forces acting on the balloon: gravity pulling the balloon downward and, because Helium gas is less dense than air, an upward buoyant force. When the balloon is first bought it is full of Helium gas that will not easily pass through the balloon and the buoyant force is much stronger than the pull of gravity. Therefore, the balloon will tend to sail into the air. As the balloon sails higher, however, the air becomes less dense, and so the buoyant force begins to decrease. At some elevation, the buoyant force and gravitational forces will exactly cancel out and the balloon will no long move up or down. The height of the balloon is the point at which two forces balance and is similar to the resting membrane voltage. As the balloon slowly leaks Helium, however, the buoyant force will decrease. The partially deflated balloon will sink to reestablish the balance of buoyant gravitational forces. Eventually, the balloon will have leaked so much Helium that it cannot overcome the force of gravity and it well fall back to the earth.

2.4.3   Definition of Resting Membrane Voltage

Although we have an intuitive feel for V_m^{rest}, we can now derive a more formal mathematical definition. At rest the membrane voltage is not changing so $\frac{dV_m}{dt} = 0$. If no stimulus is being applied, then Eq. (2.8) reveals that I_{ion} must equal zero. The meaning is that no *net* current is crossing the membrane. We can only say net current because in general, I_{ion} may be composed of many currents which may balance one another.

As F_ϕ and F_C are the driving forces behind ion movement, if $F_C \neq F_\phi$ then charged particles will cross the membrane, i.e., $I_{\text{ion}} \neq 0$. For $I_{\text{ion}} = 0$ this means that $F_C = F_\phi$, or alternatively, the current due to the potential gradient, I_ϕ, is equal to the current due to the concentration gradient, I_c. In the next sections we will use Fick's Law and Ohm's law to define I_ϕ and I_c.

2.4.4   Fick's Law and Chemical Gradients

Fick's Law describes the *flux* ($\frac{\text{mol}}{s \cdot m^2}$) of ions through an area of membrane with a thickness of dx due to a concentration gradient

$$I_c = -D\frac{dC}{dx} \tag{2.24}$$

where the *diffusion coefficient*, D ($\frac{m^2}{s}$), is a material property of the membrane and is a measure of how easily ions can pass. C is a concentration in $\frac{\text{mol}}{m^3}$.

2.4.5   Ohm's Law and Electrical Gradients

As stated earlier, Ohm's Law for a passive membrane is

$$V_m = I_\phi R_m \tag{2.25}$$

$$I_\phi = \frac{V_m}{R_m} = \frac{\phi_i - \phi_e}{R_m}. \tag{2.26}$$

The specific membrane resistivity, R_m, may be thought of as the ability of some ion to pass across the thickness (dx) of the membrane

$$R_m = \frac{dx}{\mu_p C} \cdot \frac{|Z|}{Z} \tag{2.27}$$

where μ_p is the ion mobility. Z is the ion valence, so $Z/|Z|$ is the charge sign (+ or -), and C is the ionic concentration.

In 1905 Einstein recognized that mobility (μ_p) and the diffusion coefficient (D) are related by

$$\mu_p = \frac{D|Z|F}{RT} \tag{2.28}$$

where R is the ideal gas constant, T is the temperature, and F is Faraday's constant. We can therefore, rewrite the equation for R_m as

$$R_m = \frac{dx \cdot RT}{DCFZ} .$$ (2.29)

Substitution back into Eq. (2.26) yields

$$I_\phi = \frac{[\phi_i - \phi_e] DCFZ}{dx \cdot RT}$$ (2.30)

$$I_\phi = \frac{DCFZ}{RT} \frac{d\phi}{dx} .$$ (2.31)

2.4.6 The Nernst Equation
Using Eqs. (2.24) and (2.31)

$$I_{ion} = I_c + I_\phi$$ (2.32)

$$I_{ion} = -D \left[\frac{dC}{dx} + \frac{CFZ}{RT} \frac{d\phi}{dx} \right] .$$ (2.33)

At rest we know that $I_{ion} = 0$

$$0 = -D \left[\frac{dC}{dx} + \frac{ZCF}{RT} \frac{d\phi}{dx} \right]$$ (2.34)

and after some rearranging

$$V_m^{rest} = E_{rest} = \frac{RT}{F} ln \left[\frac{C_e}{C_i} \right] .$$ (2.35)

Equation (2.35) is the *Nernst* equation that relates V_m^{rest} to the difference in intracellular and extracellular concentrations. In our derivation we assumed that C was a positive ion. The only difference in Eq. (2.35) for negative ions is that C_e and C_i are switched in the numerator and denominator. Figure 2.8 is the circuit analog for the passive membrane where the Nernst potential is represented by a battery, E_{rest}.

The ionic current that flows through the resistor is described by

$$I_{ion} = \frac{1}{R_m} [V_m - E_{rest}] .$$ (2.36)

It can now be understood why the membrane settles to $V_m^{rest} = E_{rest}$. If $V_m > V_m^{rest}$, then I_{ion} will be positive. If $V_m < V_m^{rest}$ then I_{ion} is negative. Therefore, E_{rest} is sometimes also referred to as the *reversal potential*.

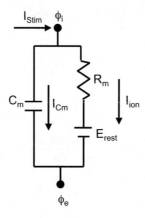

Figure 2.8: Passive membrane circuit analog with a resting potential.

2.4.7   The Goldman-Hodgkin-Katz Equation

In reality, there are many ions with different abilities to cross the membrane. There are also some large charged particles (e.g., proteins) that cannot cross the membrane. The result is that achieving $I_{ion} = 0$ to calculate V_m^{rest} is somewhat more difficult. We will not provide a derivation here but as the major ions involved are Sodium, Potassium, and Chloride, V_m^{rest} can be approximated by the Goldman-Hodgkin-Katz equation:

$$E_m = \frac{RT}{F} ln \left[\frac{P_K[K^+]_e + P_{Na}[Na^+]_e + P_{Cl}[Cl^-]_i}{P_K[K^+]_i + P_{Na}[Na^+]_i + P_{Cl}[Cl^-]_e} \right] \tag{2.37}$$

where the P-terms are the relative *permeabilities* (unitless) of the respective ions.

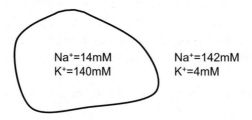

Figure 2.9: Multiple ions species and the Goldman-Hodgkin-Katz equation.

Figure 2.9 shows a simple example for only Na^+ and K^+ ions when the cell is at rest. Given that $P_K = 1$, $P_{Na} = 0.002$, $R = 8.316 \frac{J}{K \, mol}$, $T = 300K$, and $F = 96487 \frac{C}{mol}$,

$$\frac{RT}{F} = 0.0258 \frac{J}{C} = 25.8 mV \tag{2.38}$$

and

$$25.8 \cdot ln \left[\frac{1 \times 4mM + 0.002 \times 142mM}{1 \times 140mM + 0.002 \times 14mM} \right] = -90mV . \tag{2.39}$$

The individual Nernst potentials for K and Na may be found using Eq. (2.35). Perhaps surprisingly the computations show that $E_{Na} = 59.8mV$ and $E_K = -91.7mV$, neither of which are exactly at -90mV. It is worth noting that V_{rest} is much closer to E_K than E_{Na} at rest. It is also important to consider that Na^+ and K^+ may still be able to cross the membrane as long as the total current, I_{ion}, sums to zero. The direction of current flow may also be found by comparing V_m^{rest} and the reversal potential for an ion. In general, a particular ion will flow in the direction that will send V_m closer to its own Nernst potential. Remember that the definition of positive current is positive charge leaving the cell.

2.5   DEVIATIONS FROM REST

It is clear that an external stimulation can cause the membrane to deviate from rest. What is not clear is how many ions must cross the membrane to have an impact on the resting voltage. Below is a derivation of how many Sodium ions must cross the membrane of a typical neuron to generate a change in membrane potential of $100mV$ over $2msec$. As the membrane voltage is simply the voltage across a capacitor

$$I_{cm} = C_m A \frac{dV_m}{dt}. \tag{2.40}$$

Where A is the surface area of the cell in cm^2. For a typical neuron, we can assume the surface area to be $3 \times 10^{-4} cm^2$. Therefore, the current required to change the membrane potential by $100mV$ in $2msec$ is

$$I_{cm} = \left[1 \frac{\mu F}{cm^2} \right] \left[3 \times 10^{-4} cm^2 \right] \frac{0.1V}{0.002sec} = 15nA. \tag{2.41}$$

Next we must find how many Sodium ions must cross the membrane to generate $15nA$ over $2msec$. Given that current is defined as one Coulomb per second

$$\left[15^{-9} \frac{C}{s} \right] [0.002s] \left[6.26 \times 10^{18} \frac{Na^+}{C} \right] = 1.878 \times 10^8 Na^+. \tag{2.42}$$

So 187 million Sodium ions must cross the membrane in $2msec$. Because the intracellular and extracellular concentrations of Sodium will affect the Sodium Nernst potential, it is important to determine how much moving these Sodium ions will change the concentrations. If we assume that the intracellular cell volume is 8pL then

$$\Delta\left[Na^+\right]_i = \frac{1.878 \times 10^8 Na^+}{\left[6.023 \times 10^{23} \frac{Na^+}{mol}\right]\left[8 \times 10^{-12}L\right]} = 4 \times 10^{-5}\frac{mol}{L} = 40\mu M \qquad (2.43)$$

As the intracellular concentration of Sodium is approximately 14mM, a change of $40\mu M$ is only a change of approximately 0.3%. As the volume of extracellular space is much larger, the percentage change in $[Na]_e$ is even smaller. The meaning of this number is that large changes in membrane voltage can be achieved with very little change in ionic concentrations.

2.6   NUMERICAL METHODS: THE EULER METHOD

In this chapter we have defined a differential equation to relate V_m to membrane currents. Analytical solutions were possible because we assumed that all terms were linear. Once terms such as I_{ion} become nonlinear (as they will in Ch. 3), analytical solutions are no longer possible. In these situations we can turn to numerical simulations using a computer. Perhaps the simplest and most used method of finding $V_m(t)$ given $\frac{dV_m}{dt}$ is the *Forward Euler* method. The general strategy is to use the current value, $V_m(t)$, and the current slope, $\frac{dV_m}{dt}$ to predict what the new membrane potential, $V_m(t + \Delta t)$, will be Δt forward in time (see Fig. 2.10). Mathematically,

$$\frac{dV_m}{dt} = \frac{1}{C_m}[-I_{ion} - I_{syn} + I_{stim}] \qquad (2.44)$$

$$\frac{\Delta V_m}{\Delta t} = \frac{1}{C_m}[-I_{ion} - I_{syn} + I_{stim}] \qquad (2.45)$$

$$\Delta V_m = \frac{\Delta t}{C_m}[-I_{ion} - I_{syn} + I_{stim}] \qquad (2.46)$$

$$V_m^{t+\Delta t} = \Delta V_m + V_m^t . \qquad (2.47)$$

Therefore, to solve a coupled set of nonlinear differential equations, we can apply Eqs. (2.46) and (2.47) over and over again to march forward in time. One word of warning is that if Δt is large, you may overshoot the real solution.

Summary

(1) The transmembrane voltage is defined as the voltage across the cell membrane.

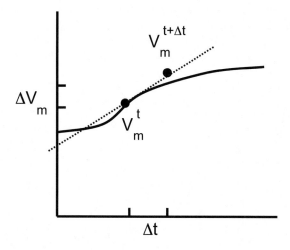

Figure 2.10: Euler method for time integration of a differential equation.

(2) The four generic transmembrane currents are due to the capacitance of the membrane, the ionic flux through ion channels, the ionic flux through synapses and any external stimuli.

(3) The relationship between transmembrane current and voltage can be made using circuit analogies. It is therefore possible to use circuit analysis to derive equations that relate changes in transmembrane voltage to transmembrane current.

(4) The strength-duration curve relates the current necessary to charge a membrane up to a predetermined voltage.

(5) The driving force for current is established by the concentrations inside and outside the cells, characterized by the Nernst potential.

(6) The resting membrane potential is the balance of Nernst potentials of many ions, characterized by the Goldman-Hodgkin-Katz Equation.

Homework Problems

(1) Assuming that $V_m^{rest} = -70mV$ and the initial voltage is $V_m = -60mV$, draw a V_m trace that repolarizes from $t = 0msec$ to $t = 30msec$, then hyperpolarizes from $t = 30msec$ to t=35msec, depolarizes from $t = 35msec$ to $t = 50msec$, and then repolarizes to rest from $t = 50msec$ to $t = 70msec$. The exact values of the peaks and valleys are not important.

(2) A negative stimulus of $-15.7pA$ was applied to a membrane at $t = 1msec$ and produced the following trace. Note that hyperpolarizing pulses are always considered to be subthreshold.

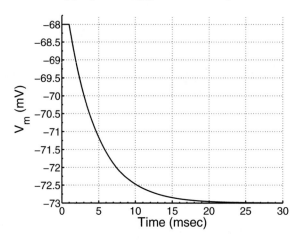

a) What is the resting membrane potential?
b) Compute the total R_m in $M\Omega$. Explain your reasoning.
c) Compute τ_m in $msec$. Explain your reasoning.
d) Compute the total cellular C_m in pF.
e) Given that the specific C_m is $1\mu F/cm^2$, find the cell surface area.

(3) The following data were obtained from a neuron at 300K (see Table 2.1).

Table 2.1:			
Ion	Intracellular (mM)	Extracellular (mM)	Permeability
$[K^+]$	280	10	0.001
$[Na^+]$	51	485	1
$[Cl^-]$	46	340	0.5

a) Report the Nernst potential for each ion.
b) Compute the resting potential.
c) Report the direction that each ionic current moves at rest.

(4) Perform the derivation from Eqs. (2.34) and (2.35).

(5) Below is a strength-duration curve for a membrane with a threshold $7mV$ above rest.

Given the figure, answer the following questions.
a) If a stimulus is above the curve, will the membrane behave linearly or nonlinearly?

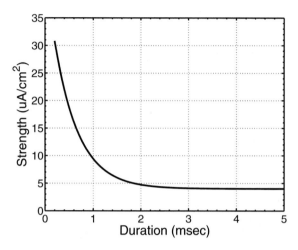

b) What is I_{rhe}?

c) What is R_m in $k\Omega cm^2$?

d) What is T_{ch} in msec?

e) What is τ_m in msec?

f) What is C_m in $\mu F/cm^2$?

(6) Draw Strength Duration curve given $C_m = 1\mu F/cm^2$, $R_m = 3k\Omega cm^2$, $V_m^{rest} = -60mV$ and $V^{th} = -50mV$. Be sure to label I_{rhe} and T_c and report numerical values for each.

(7) Demonstrate that the units in Eq. (2.11) are the same on the left- and right-hand side of the equation.

(8) Cellular currents are often measured in units of $\mu A/pF$. Explain why this may be a good measure to use and explain how you would convert Eq. (2.11) to the proper form.

Simulation Problems

(1) Program a passive membrane with input variables of Rm, Cm, dt, endtime, Stim start, Stim End, and Stim Strength.

CHAPTER 3

Active Membranes

In the previous chapter we considered the response of a membrane to a small stimulus. In these situations the resistance of the membrane was linear and I_{ion} was modeled as a simple resistor and battery in series. In this chapter, we consider what happens when a stimulus causes V_m to reach threshold. The result is that R_m no longer behaves linearly. The nonlinearity of the membrane is represented in a circuit model as the *variable resistance* in Fig. 3.1.

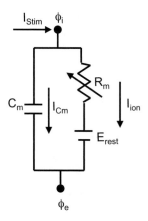

Figure 3.1: A nonlinear resistive membrane.

3.1 THE HODGKIN-HUXLEY MODEL

The first physiologically accurate nonlinear model of I_{ion} was published in 1952 by Hodgkin and Huxley. To create the model, Hodgkin and Huxley combined a brilliant experiment performed on a giant squid axon and a number of assumptions about the biophysics of ion channels. Despite its relative simplicity, it remains the gold-standard of ionic membrane models for all excitable cells.

3.1.1 The Parallel Conductance Model
The first assumption was that I_{ion} was composed of three currents that acted independently of one another. These currents were Sodium (I_{Na}), Potassium (I_K), and a generic *leakage* current (I_L). Mathematically,

$$I_m = C_m \frac{dV_m}{dt} + I_{ion} \qquad (3.1)$$
$$I_{ion} = I_{Na} + I_K + I_L . \qquad (3.2)$$

Experimental data showed that I_L was a linear current, while I_{Na} and I_K were nonlinear. Therefore, the circuit analog, taking into account the independence assumption and nonlinear ionic currents, can be represented as in Fig. 3.2. The resistors have been replaced by G-terms that represent *conductances*. Conductance is simply the inverse of resistance ($G = \frac{1}{R}$) and is measured in units of *Siemens*, abbreviated with a capital 'S'. Therefore, as resistance is increased, conductance is decreased. For biological membranes the most commonly used unit for conductance is mS/cm^2.

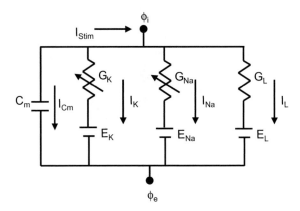

Figure 3.2: Hodgkin-Huxley circuit analog.

3.1.2   The Leakage Current
The leakage current is linear and so can be formulated in the same way as the linear current in Eq. (2.36).

$$I_L = G_L[V_m - E_L] \qquad (3.3)$$

where G_L is the *leakage conductance* and E_L is the leakage Nernst potential. The nature of the leakage current was not fully known to Hodgkin and Huxley but they guessed (again correctly) that it was some combination of other ionic currents.

3.1.3   Nonlinear Currents and the Voltage Clamp
Characterization of the two nonlinear currents (I_{Na} and I_K) is nontrivial because there is a natural feedback loop between I_{ion} and V_m. Consider the following generic nonlinear current

$$I_{nl} = G_{nl}(V_m)[V_m - E_{nl}] . \tag{3.4}$$

Notice that the conductance, G_{nl}, is a function of V_m. Therefore, a change in V_m causes a change in G_{nl} which has an impact on I_{nl}. As I_{nl} is a component of I_{ion}, any nonzero value for I_{nl} may in fact lead to a further change in V_m. Untangling this interdependence in an experiment was achieved by a technique called the *voltage clamp*. It was developed by Cole and was a critical step in the mathematical characterization of I_{Na} and I_K.

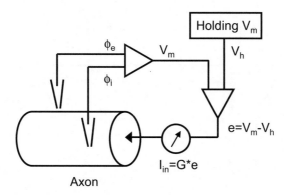

Figure 3.3: Voltage clamp circuit.

The voltage clamp is a method of forcing V_m to be constant at any desired holding voltage, V_h, using an external circuit (Fig. 3.3). In this way, the feedback loop between V_m and I_m is effectively cut. If the circuit can react faster than the membrane, an exactly counterbalancing current, I_{in}, may be sent into the cell to maintain a specific *holding voltage*. The important data to be gained from this experiment is the time trace of I_{in} as the holding voltage is quickly changed to a different holding voltage level. The transient of this counterbalancing current can be used to determine how fast the membrane can react to changes when $V_m = V_h$, while the steady-state current can be used to determine the maximum current at the clamped V_m. The combination of these two parameters, reaction time and maximum current, as a function of V_h, enabled Hodgkin and Huxley to develop the first mathematical model of an excitable cell.

It is clear from Fig. 3.3 that at least one electrode must have access to the inside of a cell. At the time of Hodgkin and Huxley, electrodes were relatively large compared to mammalian neurons. To create their model, Hodgkin and Huxley chose to study the large axon (up to $1mm$ in diameter) of the giant squid. Since 1952, the size of electrodes has drastically shrunk, allowing for many types of neurons to be studied. The electronic feedback loop of the voltage clamp, however, has remained largely unchanged.

3.1.4   The Sodium Current

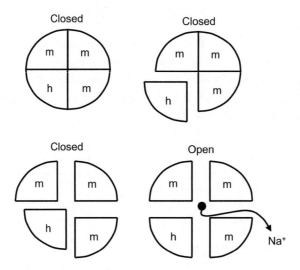

Figure 3.4: Schematic of Sodium channel dynamics.

The Sodium current is formulated in the same way as the leakage current.

$$I_{Na} = G_{Na}(V_m)[V_m - E_{Na}] \tag{3.5}$$

but now the leakage term G_{Na} is a nonlinear function of V_m. To begin, Hodgkin and Huxley assumed (again correctly) that the nonlinear nature of the flow of Sodium ions was due to the opening and closing of proteins that spanned the membrane. These proteins could somehow change their shape to be either open or closed in response to pH, intra and extracellular concentrations of ions and molecules, temperature, or even V_m. Hodgkin and Huxley focused on the impact of changes in the membrane voltage. Yet another assumption was that each ion channel would be either entirely open or entirely closed. In other words, there would be no in between states. To create their model, however, they assumed that there were thousands or millions of Sodium channels embedded in a small patch of membrane. Therefore, the model would be based on the *probability* that any individual channel would be open. So, although each ion channel would be either open or closed, 30% of the channels may be open and 70% are closed. To capture the aggregate behavior, Hodgkin and Huxley defined a variable, O, that would be 0.3, or the probability of randomly picking an open channel. Next they assumed the rate at which channels opened (on average) was not necessarily the same as the rate at which channels closed. This situation can be represented kinematically by

$$O \underset{\beta(V_m)}{\overset{\alpha(V_m)}{\rightleftharpoons}} C$$

where rate constants, $\alpha(V_m)$ and $\beta(V_m)$, would be functions of V_m. Using this formulation they could write down a differential equation to describe the dynamics of the O variable.

$$\frac{dO}{dt} = \alpha(V_m)(1 - O) - \beta O(V_m) . \tag{3.6}$$

Given the variable O, they defined the Sodium current as:

$$I_{Na} = g_{Na} \times O(V_m)[V_m - E_{Na}] \tag{3.7}$$

where g_{na} is the *maximum conductance*. The meaning of g_{na} is that if $O = 1$, i.e., every sodium channel is open, the maximum possible current will flow. O, on the other hand, can vary between a probability of 0 and 1. When Hodgkin and Huxley fit parameters for α and β using the voltage clamp, they found that the channel was more complicated than the simple O variable could capture. They assumed that the Sodium channel was composed of four parts (Fig. 3.4) and for the channel to be fully open, all four parts needed to be in the right configuration. They therefore assumed I_{Na} to take the form of:

$$I_{Na} = g_{Na}O_{1(V_m)}O_2(V_m)O_3(V_m)O_4(V_m)[V_m - E_{Na}] \tag{3.8}$$

where $O_{1,2,3,4}$ were the four parts. Further analysis showed that three of the parts functioned in the same way, i.e., α and β functions were identical. These three channel parts were each called m and the remaining part was called h. Their model of the Sodium current was described by

$$I_{Na} = g_{Na}m^3h[V_m - E_{Na}] \tag{3.9}$$
$$\frac{dm}{dt} = \alpha_m(1 - m) - \beta_m m \tag{3.10}$$
$$\frac{dh}{dt} = \alpha_h(1 - h) - \beta_h h \tag{3.11}$$

where m and h are known as *gating* variable because they control how ions are gated through the channel. The only difference between the h and m variables is that the functions of α and β are different. Using their experimental data and some curve fitting, Hodgkin and Huxley found the following fits for the α's an β's:

$$\alpha_m = 0.1 \frac{25 - v_m}{e^{(25-v_m)/10} - 1} \tag{3.12}$$

$$\beta m = 4e^{-v_m/18} \tag{3.13}$$

$$\alpha_h = 0.07 e^{v_m/20} \tag{3.14}$$

$$\beta_h = \frac{1}{e^{(30-v_m)/10} + 1} \tag{3.15}$$

where v_m is scaled by the resting potential such that $v_m = V_m - V_m^{\text{rest}}$.

3.1.5   The Potassium Current

The Potassium current may be assumed to be of the nonlinear form

$$I_K = g_K O(V_m)[V_m - E_K] . \tag{3.16}$$

Recall, however, that the first assumption was that each current was independent of the other currents present. So, simply performing the voltage clamp as described above would yield data on $I_K + I_{Na}$. To separate the two currents, Hodgkin and Huxley used tetrodotoxin (TTX), a Sodium channel blocker, to isolate I_K. They then performed the same experiment without TTX to yield $I_K + I_{Na}$ and by simple subtraction they isolated I_{Na}. Using this two-step procedure, they found that $O(V_m)$ for I_K was the product of four gating variables that were all identical:

$$I_K = g_K n^4 [V_m - E_K] \tag{3.17}$$

$$\frac{dn}{dt} = \alpha_n(1 - n) - \beta_n n . \tag{3.18}$$

Experimental data for the n gating variable was fit to the following α and β functions:

$$\alpha_n = 0.01 \frac{10 - v_m}{e^{(10-v_m)/10} - 1} \tag{3.19}$$

$$\beta_n = 0.125 e^{-v_m/80} . \tag{3.20}$$

3.1.6   Steady-State and Time Constants

An alternative, and possibly more intuitive, way of writing the gating differential equations is

$$\frac{dm}{dt} = \frac{m_\infty - m}{\tau_m} \tag{3.21}$$

$$m_\infty = \frac{\alpha_m}{\alpha_m + \beta_m} \tag{3.22}$$

$$\tau_m = \frac{1}{\alpha_m + \beta_m} \tag{3.23}$$

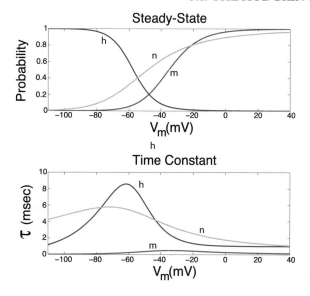

Figure 3.5: Steady-state and time constant for Hodgkin-Huxley gates.

with similar equations to describe the h and n variables. Note that τ_m in Eq. 3.23 has a different meaning than the membrane time constant in Chapter 2. Plots of the steady-state and time constant curves for m (red), h (blue), and n (green) are shown in Fig. 3.5. The reason for writing the equations this way is that the steady state (∞-terms) and time constants (τ-terms) have a physical interpretation. The solution to Eq. (3.21) is

$$m(t) = m_\infty - (m_\infty - m_0)e^{-t/\tau_m} \tag{3.24}$$

where m_0 is the initial value of m. In the context of the voltage clamp, consider that V_m is equal to a value A and has been there for a long time, i.e., $t \to \infty$. Therefore, $m \to m_\infty(A)$ which could be read directly from Fig. 3.5 Next, consider that the membrane potential is suddenly changed to $V_m = B$. Since the gating variable, m, cannot change instantaneously, the initial condition m_0 is equal to $m_\infty(A)$. The steady-state and time constants, however, do change instantaneously. Therefore, to compute $m(t)$ at any point after V_m is clamped to B

$$m(t) = m_\infty(B) - (m_\infty(B) - m_0)e^{-t/\tau_m(B)} \tag{3.25}$$
$$m(t) = m_\infty(B) - (m_\infty(B) - m_\infty(A))e^{-t/\tau_m(B)} . \tag{3.26}$$

The intuitive interpretation is that m_∞ is the target for m and τ_m is the rate at which this new target is reached. To demonstrate this idea, Fig. 3.6 shows three changes in V_h with the corresponding change in m. It may be helpful to think about how Figs. 3.5 and 3.6 are related.

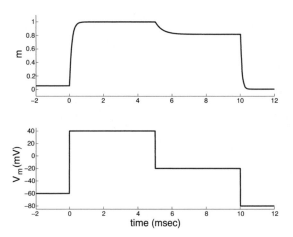

Figure 3.6: m-gate changes over time for changes in holding potential.

3.1.7   A Game of Tag

An analogy to the steady states and time constants may be thought of as a game on an obstacle course. You and 100 of your friends all have the goal of reaching a diamond that will be set somewhere on the course. You can imagine that if the diamond stays in one location, you would simply navigate through water, hoops, sand, and gravel to get to it. Although this may take time, eventually you would all reach the diamond. The location of the diamond therefore is analogous to the steady state. If the location of the diamond, however, suddenly changed, you and your friends would again set out across the course to reach it. The type of obstacles in your way (not necessarily the distance) would determine how long it takes to reach the diamond. The way that your time to reach the diamond is dependent upon the obstacles in your way is analogous to the time constant. Now imagine that the diamond is moving around the course, suddenly appearing in one location and then quickly moving to a new location, faster than you and your friends can keep up.

Next imagine that at the same time that you and your friends are chasing a diamond, another group of 101 people are also chasing a ruby, and a third set of 101 people are chasing an emerald. Here, the analogy is that the diamond is the m-gate, the ruby is the h-gate, and the emerald is the n-gate. Although our game is not a perfect analogy to the dynamics of gates, it may help understand the next section where the Hodgkin-Huxley action potential is examined.

3.2   THE HODGKIN-HUXLEY ACTION POTENTIAL

When stimulated, the Hodgkin-Huxley model generates a rapid depolarization followed by a slower repolarization back to rest. This cycle is known as an *action potential* and is the basic functional unit of all excitable cells. Below we discuss the four phases of the action potential as summarized in the V_m panel of Fig. 3.7.

3.2.1   Phase 0 - Rest

When V_m is at rest $(-60mV)$, $m = m_\infty \approx 0.05$ and $h = h_\infty \approx 0.6$ and therefore $g_{na}m^3h$ is small (See Fig. 3.5). The result is that little I_{Na} will flow. On the other hand, V_m^{rest} is close to E_K, so the driving force for I_K is small. For these different reasons, both I_{Na} and I_K are small. The result is that at rest the linear leakage current, I_L, will dominate and the membrane behaves linearly.

3.2.2   Phase 1 - Activation

If a stimulus is applied, V_m will depolarize. As V_m changes, the α, β, time constants and steady-state values will also change. Consider the following:

1. The driving force for I_{Na} (e.g., $V_m - E_{Na}$) at rest is large and negative because V_m is much smaller than E_{Na}. If given the chance, Na^+ ions would therefore move *into* the cell.

2. Na^+ ions are prevented from crossing the membrane because m is small. Physically the Na^+ channels are closed.

3. Although the m_∞ curve is flat at V_m^{rest}, if the membrane is depolarized slightly, m_∞ will become relatively large. This is because of the steep slope of the m_∞ curve.

4. τ_m is small so any change to m_∞ will result in a fast change in m.

5. The driving force for I_K at rest is small because V_m is close to E_K.

Given these properties, consider that a small depolarizing change in V_m would cause m_∞, and therefore m, to increase quickly. The result is that $g_{Na}m^3h$ and thus I_{Na} increase and Na^+ will rush into the cell. As Na^+ rushes into the small volume of the cell, the intracellular potential, ϕ_i, will increase. The extracellular space, however, is large and so ϕ_e will change only slightly. The result is that the cell membrane will become more depolarized.

The fast rise in the membrane potential is called *activation* and the m gate is therefore called the *activation* gate. The origin of a threshold voltage can now be understood. Below V_m^{th}, m_∞ and m are not large enough to create a large I_{Na}. But, when V_m reaches V_m^{th}, m is on the steep part the m_∞ curve. So, a small depolarizing I_{Na} current pushes V_m and m to larger and larger values. This runaway effect is counterbalanced by *repolarization*.

3.2.3   Phase 2 - Repolarization

As the I_{Na} current attempts to drive $V_m \rightarrow E_{Na}$, two other players become important. First, during the upstroke h_∞ changes from ≈ 0.6 to ≈ 0, but does so slowly, i.e., τ_h is relatively large compared to τ_m. Over time, however, h becomes small enough that the m^3h term, and therefore I_{Na}, is reduced. The h gating variable is therefore called the *inactivation* gate because it turns the Sodium current off after activation. Second, after depolarization, V_m is no longer close to E_K and the driving force for I_K is increased. Due to the sign of E_K, I_K will be a positive current, i.e., K^+ flows *out* of the cell. Positive charge exiting the cell causes V_m to decrease or *repolarize*.

3.2.4   Phase 3 - Hyperpolarization

In the Hodgkin-Huxley model, the *n*-gate does not return the membrane directly to rest. Instead, V_m drops slightly below V_m^{rest}, a *hyperpolarization*, and then depolarizes back to rest.

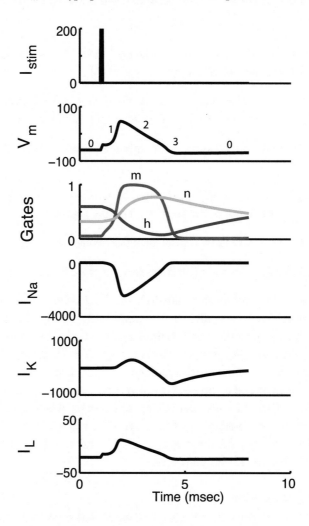

Figure 3.7: Hodgkin-Huxley action potential and currents.

3.3 PROPERTIES OF NEURONAL ACTION POTENTIALS

The Hodgkin-Huxely model was successful largely because it explained the origin of four of the most important features of an action potential. We review these properties below and related them to the relevant parts of the model.

3.3.1 All Or None

When a stimulus of a small amplitude is applied to an active membrane, the membrane will act in the same way as a passive membrane. We argued in Ch. 2 that at some combination of strength and duration (recall Fig. 2.6) the membrane will reach a threshold, V^{th}, after which the membrane will become nonlinear. Furthermore, the Hodgkin-Huxely model correctly predicts that once the membrane is above threshold, the Sodium current begins the positive feedback cycle that results in the upstroke of the action potential. Therefore, once the Sodium current is activated, an action potential is inevitable. This abrupt *all or none* response also highlights the importance of the strength-duration curve because any stimulus above the curve will result in an action potential.

3.3.2 Refractory Periods

After I_{Na} has caused the upstroke of the action potential and repolarization has begun, m will remain relatively high. The h-gate, however, will slowly drop to a value forcing I_{Na} to become small. But remember that τ_h is large so any changes in h occur slowly. Therefore, h will remain small for some time while the repolarization of the action potential is occurring. During this time, no additional change in m will cause a second large influx of I_{Na}. The time when no second action potential is possible is called the *absolute refractory period* (see Fig. 3.8).

Following the absolute refractory period is the *relative refractory period*. During the relative refractory period a second action potential is possible, but requires a greater depolarization (e.g., larger I_{stim}) than the first action potential.

3.3.3 Anode Break

An interesting phenomenon can occur if a hyperpolarizing current is applied to the membrane for a long time. In this case, m is forced low and because the current is on for a long time, h is forced high. When the stimulus is turned off, m will respond quickly and increase. h, on the other hand, will respond slowly and stay at nearly 1. The combination of these two factors can cause $g_{Na}m^3h$ to reach a level high enough for I_{Na} to increase in magnitude. As in a normal activation, the increase in I_{Na} can lead to the run-away process that causes the upstroke. The firing of an action potential using a long duration hyperpolarizing current is known as *anode break* and is shown in Fig. 3.9.

3.3.4 Accommodation

Another interesting phenomenon can be observed if the membrane is very slowly depolarized. A slow depolarization can be accomplished by making I_{stim} a ramp with a small slope as in Fig. 3.10. If the depolarization is slow enough, the h gate will have time to reach h_∞ (unlike in a normal

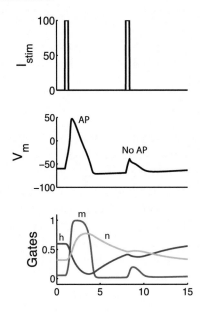

Figure 3.8: Demonstration of action potential refractory period.

depolarization). Therefore, even though m increases, h will decrease such that $g_{Na}m^3h$ will not be large enough to initiate the run-away I_{Na} current. The effect is that V_m can become higher than V_m^{th} without causing an action potential to fire. This phenomenon is called *accommodation* because the h gate is impacting the value of the threshold. The meaning of both anode break and accommodation is that what we typically think of as the threshold voltage is not an exact number and depends on the path taken to cause the run-away I_{Na}. Both are related to the time constant of h.

3.4   COMPLEX IONIC MODELS

While Hodgkin and Huxley created the first physiologically based mathematic model of a neuron, there were many neuronal phenomenon that it could not reproduce. Below we review some of the ways researchers have extended the Hodgkin-Huxley model.

3.4.1   More Currents

As the electrodes used for the voltage clamp method became smaller and smaller, and new channel blocking drugs were found, it became possible to isolate more than the three simple currents of the Hodgkin-Huxely model. The result has been that more recent models have many more currents included in I_{ion}. For example, the unspecified leakage current was found to be due to other ions such as Ca^{2+}, Mg^{2+}, and Cl^-. To include these currents we would simply add them to the parallel conductance I_{ion} term.

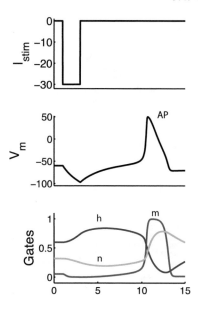

Figure 3.9: Demonstration of an action potential generated by anode break.

$$I_{\text{ion}} = I_{Na} + I_K + I_{Ca} + I_{Mg} + I_{Cl} . \tag{3.27}$$

Here the leakage current has been replaced by $I_{Ca} + I_{Mg} + I_{Cl}$. It was also found that I_K may in fact be the summation of many different types of Potassium ion channels that act independently.

$$I_{\text{ion}} = I_{Na} + I_{K1} + I_{K2} + I_{K3} + I_{Ca} + I_{Mg} + I_{Cl} \tag{3.28}$$

where I_K has been split into three different currents. Similarly, I_{Ca} may be decomposed into many currents. For each current added, there will be new gating variables and new differential equations.

It may be questioned why the addition of these currents is necessary. After all, the Hodgkin–Huxely model successfully reproduced the main properties of action potentials. There are at least two reasons why the extra currents and complexity are important. First, the giant squid axon does not display all of the behavior of mammalian neurons. Second, it is often the case that a genetic mutation, drug, or hormone will only effect one type of ion channel. For example, Nifedipine specifically blocks a certain type of Calcium ion channel, while allowing other Calcium channels to remain functional. Second, when a single action potential fires, Na^+, K^+, and other ions cross the membrane. Surprisingly, it does not require many ions to cross the membrane to achieve the $\approx 100mV$ change in V_m during the action potential. Therefore, the intracellular and extracellular

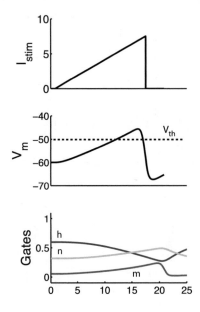

Figure 3.10: Demonstration of accommodation of threshold voltage.

concentrations do not change significantly during a single action potential. If many action potentials fire in rapid succession, however, the concentrations will eventually begin to change. Cells therefore require a variety of *pumps* and *exchangers* that slowly act in the background, transporting ions back across the membrane. As these pumps are typically working against a concentration gradient, they require ATP, and therefore a blood supply that can supply Oxygen. Similar to other currents, they may be added to I_{ion}. We will consider how to incorporate changing concentrations in Sec. 3.4.5.

3.4.2   The Traub Model of the Pyramidal Neuron

A example of model with additional currents is the Traub model. We present the model here as an example of how extra currents lead to more realistic behavior and because it will be encountered again in Ch. 7.

$$I_{ion} = I_{Na} + I_K + I_{Ca} + I_{kCa}$$
$$I_{Na} = g_{na}m^3h[V_m - E_{Na}]$$
$$I_K = g_k n^4 y[V_m - E_K]$$
$$I_{Ca} = g_{Ca}s^5 r[V_m - E_{Ca}]$$
$$I_{kCa} = g_{kCa}q[V_m - V_k]$$

where the gates are defined in Table 3.1.

gate	α	β
Table 3.1:		
s	$\dfrac{0.04(60-V_m)}{\exp\left(\frac{60-V_m}{10}\right)-1}$	$\dfrac{0.005(V_m-45)}{\exp\left(\frac{V_m-45}{10}\right)-1}$
m	$\dfrac{0.32(12-V_m)}{\exp\left(\frac{12-V_m}{4}\right)-1}$	$\dfrac{0.28(V_m-40)}{\exp\left(\frac{V_m-40}{5}\right)-1}$
h	$0.128\exp\left(\dfrac{17-V_m}{18}\right)$	$\dfrac{4}{\exp\left(\frac{20-V_m}{5}\right)+1}$
n	$\dfrac{0.032(15-V_m)}{\exp\left(\frac{15-V_m}{5}\right)-1}$	$0.5\exp\left(\dfrac{10-V_m}{40}\right)$
y	$0.028\exp\left(\dfrac{15-V_m}{15}\right)+\dfrac{2}{\exp\left(\frac{85-V_m}{10}\right)+1}$	$\dfrac{0.4}{\exp\left(\frac{40-V_m}{10}\right)+1}$
r	0.005	$\dfrac{0.025(200-x)}{\exp\left(\frac{200-x}{20}\right)-1}$
q	$\dfrac{0.005(200-x)\exp\left(\frac{V_m}{20}\right)}{\exp\left(\frac{200-x}{20}\right)-1}$	0.002

and the differential equations for each gate are of the form shown in Eq. (3.6). One additional x-gate is defined by

$$\frac{dx}{dt} - \frac{c \cdot I_{Ca}}{A \cdot d} - \beta_x \cdot x \ .$$

The constants for the Traub model appear in Table 3.2.

Table 3.2:	
$\beta_x = 0.1$	
$A = 3320$	$c = 5.2$
$d = 0.0005$	$E_{Ca} = 140mV$
$E_{Na} = 115mV$	$E_k = -15mV$
$g_{Na} = 3.32\mu S$	$g_{kCa} = 0.1\mu S$
$g_{Ca} = 6.64\mu S$	$g_K = 3.98\mu S$

The additional phenomena the Traub model simulates is a repeating cycle of depolarization and repolarization called *bursting*. Bursting is often modulated by the strength of a sustained input current and is shown in Fig. 3.11.

3.4.3 Complex Gating

Hodgkin and Huxley assumed that ion gates could be either open or closed. But, the h gate was a clue to the more complex nature of channels. In reality, some channels may be open in more than

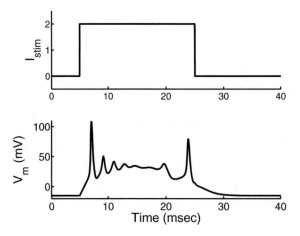

Figure 3.11: Bursting in the Traub membrane model.

one way, closed in more than one way, and even inactivated in more than one way. The method of modeling the complex dynamics of an ion channel is through a *Markov State* model. Figure 3.12 is an example of the more complex dynamics of a Sodium channel. Each circle is a state that the channel may be in. The α and β terms are the rates of transition from one state to another and typically do not depend upon V_m. To derive the differential equations for a state, we subtract the sum of the state and all outflowing rates (exiting arrows) from the product of the entering states and inflow rates (entering arrows). For example, the equations for the model in Fig. 3.12 are:

$$\frac{dC3}{dt} = \beta_{23}C2 - \alpha_{32}C3 \tag{3.29}$$

$$\frac{dC2}{dt} = \beta_{12}C1 + \alpha_{32}C3 - (\alpha_{21} + \beta_{23})C2 \tag{3.30}$$

$$\frac{dC1}{dt} = \beta_{01}O + \alpha_{21}C2 - (\alpha_{10} + \beta_{12})C1 \tag{3.31}$$

$$\frac{dIF}{dt} = \beta_{45}IS + \alpha_{40}O - (\alpha_{54} + \beta_{40})IF \tag{3.32}$$

$$\frac{dIS}{dt} = \alpha_{54}IF - \beta_{45}IS \tag{3.33}$$

$$\frac{dO}{dt} = \beta_{40}IF + \alpha_{10}C1 - (\alpha_{40} + \beta_{10})O . \tag{3.34}$$

The physical interpretation of complex gating is that an ion channel is more than a simple tunnel. There are sections that are energetically narrow and therefore require more energy to traverse. Other sections of the channel enable easier passage. These barriers to ionic flow are represented by the α and β coefficients. The advantage of this type of model is that there is a more direct link between the genetic code, expressed protein, structure of the channel, and electrophysiologic function. It is hoped that someday we will be able to engineer sections of genetic code and simulate their function before they are made. In this way, diseased channels could be repaired by making only the correction that is needed. Such a simulation of the function will require a Markov model.

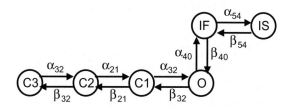

Figure 3.12: Markov model of Sodium channel.

3.4.4   Gating Dependence on pH, Temperature and Concentrations

It was mentioned in Sec. 3.1.4 that ion channels are often gated by V_m but could also be gated by pH, ionic or molecular concentrations, temperature, or some combination of all of these gating mechanisms. For example, the functions for α and β in the Hodgkin-Huxely model were for a particular temperature. In general, these functions may be much more complex, for example $\alpha \left[V_m, pH[Mg^{2+}]_e, [Atropine], T \right]$. Although this may seem to be a daunting modeling task, there are some very important ion channels that are not gated by V_m and are only sensitive to the concentration of some ion or molecule. These channels are said to be *Ligand*-gated and typically play an important role in mutations, the impact of drugs and the onset of diseases. In Ch. 6, the dynamics of the synapse will be modeled as gating by concentration of neurotransmitter.

3.4.5   Changes in Nernst Potentials

In section 2.5 the percentage change in $[Na]_i$ was shown to be small for a transmembrane voltage change of $100 msec$ over $2 msec$ (on the order of a single action potential upstroke). Even with pumps and exchangers working in the background, the cell may build up concentrations of some ions. Accompanying these concentration changes, will be a change in the Nernst and resting potentials, and thus the driving forces behind the action potential. It is therefore important for some models to incorporate a way for concentrations to change. The most common technique is to write differential equations for the intra and extracellular ionic concentrations. For example, consider that there may be several I_K outward currents that could potentially deplete the intracellular Potassium concentration, $[K^+]_i$. This change in $[K^+]_i$ would impact E_K as well as V_m^{rest} through the Goldman-Hodgkin-Katz equation. A common formulation would be:

$$\frac{d[K^+]_i}{dt} = -\frac{I_{K1} + I_{K2} + I_{K3}}{FV_i} \tag{3.35}$$

$$\frac{d[K^+]_e}{dt} = \frac{I_{K1} + I_{K2} + I_{K3}}{FV_s} \tag{3.36}$$

where F is Faraday's constant, V_i is the volume of the intracellular space, and V_s is the small *shell* of extracellular space surrounding the outside of the cell. In this formulation, any K^+ ions leaving the intracellular space must enter the shell of extracellular space around the cell.

A situation where ionic concentrations may drastically change is during a disease. For example, during ischemia (e.g., lack of O_2) the concentration of ATP drops and the pumps that restore concentration gradients become less efficient and may even fail. The result is a significant change in intracellular or extracellular ionic concentrations. If this buildup continues, cells may die and lyse (e.g., pop) spilling their ions into the extracellular space and lead to a drastic change in concentrations.

3.4.6   Intracellular Compartments and Buffers

The intracellular space in a cell is populated by many smaller organelles. Some of these organelles have their own membranes, with their own ion channels, and are capable of transporting ions to and from the intracellular space to the intra-organelle space. When this transport occurs, the concentration of that ion will change in the intracellular space. Functionally, these organelles behave as a *buffer*. As in the previous section, these changes in concentration can impact the Nernst potential of that ion. Fig. 3.13 is a schematic of a Ca^{2+} buffer. These buffer currents can be treated in the same way as the transmembrane currents. Ionic concentrations can be tracked using the equations in Section 3.4.5.

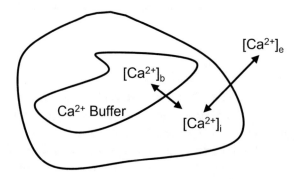

Figure 3.13: Ionic buffering by intracellular compartments.

3.5 PHENOMENOLOGICAL MODELS

Although the Hodgkin-Huxley mathematical model is simple to solve using todays computers, Hodgkin and Huxley performed all of their calculations using calculators. In particular, computing exponentials was difficult, so a number of simplified models were developed that captured the basic features of neuronal action potentials.

3.5.1 Fitzhugh-Naghumo Model

In 1961, Fitzhugh and Nagumo independently developed a model based upon simple polynomials. In the Fitzhugh-Naghumo model, the m, h, n, and V_m variables were reduced to only two differential equations.

$$\frac{dV_m}{dt} = V_m - \frac{V_m^3}{3} - W + I_{\text{stim}} \tag{3.37}$$

$$\frac{dW}{dt} = a\left[V_m + b - cW\right] . \tag{3.38}$$

The model has a steady-state resting potential, fast all-or-none upstroke, and slower repolarization. Furthermore, the parameters a, b, and c can be tuned to generate action potentials of different shapes and durations.

3.5.2 Hindmarsh-Rose Model

One disadvantage of the Fitzhugh-Naghumo model is that it does not generate bursting behavior. In 1984, Hindmarsh and Rose developed a set of three differential equations that would allow for the phenomenon of bursting:

$$\frac{dV_m}{dt} = -aV_m^3 + bV_m^2 + y - z + I \tag{3.39}$$

$$\frac{dy}{dt} = -dV_m^2 - y + c \tag{3.40}$$

$$\frac{dz}{dt} = rsV_m - rz - rsV_{\text{rest}} \tag{3.41}$$

where a, b, c, d, r, s, and V_{rest} are constants and may be tuned to producing different bursting behavior.

3.5.3 Integrate and Fire Model

In 1907, long before Hodgkin and Huxley, Lapicque proposed that the firing of an action potential could be modeled simply as a spike in voltage. As I_{stim} is applied, V_m will depolarize according to the familiar passive model

$$\frac{dV_m}{dt} = \frac{1}{C_m}\left[-G_L(V_m - E_L) + I_{\text{stim}}\right] . \tag{3.42}$$

When V_m reaches some predefined V_m^{th}, however, all solving of differential equations is suspended. The simulation then abruptly jumps V_m to some value, V_m^{peak}, to simulate the upstroke of the action potential. In some implementations, V_m is clamped to V_m^{peak} for some short time and then *reset* back to the resting value as in Fig. 3.14. In other implementations, the reset is not exactly back to rest but below V_m^{th}. This simple yet elegant model is called the *integrate and fire* model. It is still used in neural modeling because it requires no updating of gating variables.

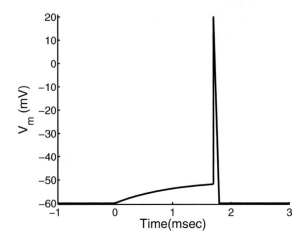

Figure 3.14: Integrate and fire action potential.

3.6   NUMERICAL METHODS: TEMPLATE FOR AN ACTIVE MEMBRANE

In Sec. 2.6, a way of numerically solving a differential equation was outlined. In the active membrane models, we need to keep track of several differential equations as well as compute rate constants, steady-state values and currents. Below is a template for how to write a program to solve the active equations.

```
Define constants (e.g., GL, gNa, gK, Cm, dt)
Compute initial αs and βs
Compute initial conditions for state variables
(e.g., Vmrest, m, h, n)

for (time=0 to time=end in increments of dt)

        Compute Currents (e.g., IL, INa, IK)
```

```
Compute αs and βs
Update Differential Equations (V_m, m, n, h) by Forward Euler
Save values of interest into an array (e.g., V_m)
```

```
end
```

```
Store values of interest to a file
```

Summary

(1) Transmembrane currents are fluxes of charged ions that cross the cell membrane through specialized protein channels. These channels operate independently of one another.

(2) The parallel conductance model provides a circuit analogy for the different pathways for ions to cross the cell membrane.

(3) The patch clamp method allows for the feedback loop between transmembrane current and voltage to be cut and therefore allows systematic experimentation on cellular membranes.

(4) The Hodgkin-Huxley model is the gold-standard model of excitable cells.

(5) Action potentials are composed of four phases; rest, activation, repolarization and in some cells, hyperpolarization.

(6) Action potentials typically display an all-or-none response to stimuli, a refractory period following activation, anode break in response to a hyperpolarizing stimuli and accommodation to slowly varying stimuli.

(7) Biophysical models of neurons can become more complex by adding additional currents to the parallel conductance model. Modeling additional currents may allow neurons to display more realistic behavior.

(8) Phenomenological models are a simplification of complex models and can provide insight into the basic mechanisms underlying the dynamics of action potentials.

Homework Problems

(1) If a stimulus is above a strength duration curve and the membrane is active, will and action potential fire? Explain.

(2) The following data were recorded at the peak of a Hodgkin-Huxely action potential:

$$I_K = 353.574 \frac{\mu A}{cm^2}$$

$I_{Na} = -394.59 \frac{\mu A}{cm^2}$

$g_K = 3.36737 \frac{mS}{cm^2}$

$g_{Na} = 130.7429 \frac{mS}{cm^2}$

$m = 0.889$

$h = 0.288$

$E_K = -75.0 mV$

$E_{Na} = 50 mV.$

a) What is the peak magnitude of the action potential in mV?

b) How does this compare to E_{Na}? What does this mean?

c) Compute the fraction of K+ channels open.

d) What is the value of I_{leak}?

(2) In a voltage-clamp experiment, the transmembrane potential (V_m) was changed from rest (-60mV) to 0mV, kept at 0mV for 100ms, and then changed to -50mV. Relevant parameters are given in the table below (time constants in msec).

Table 3.3:				
V_m	α_m	τ_m	α_h	τ_h
-60mV	0.225	0.238	0.0697	8.51
-50mV	0.433	0.368	0.0423	6.16
0mV	3.62	0.266	0.00347	1.05

a) Sketch the holding potential and label the times and voltage levels.

b) Compute the fraction of h gates open at $t = 100 msec$.

c) Compute the fraction of h gates that are open at $t = 104 msec$.

d) Compute the probability of an Na+ channel being open at $t = 104 msec$.

(3) Extend the Hodgkin-Huxely model to include Potassium buffering by an intracellular organelle. Be sure to write down differential equations for the concentrations of K^+ in the organelle, cell, and extracellular space. Include two additional currents, one between intra and extracellular space and one between intracellular and organelle space.

(4) Write down the differential equations to describe the Markov Model of the K^+ channel shown below.

(5) Show how the units work for Eq. 3.35.

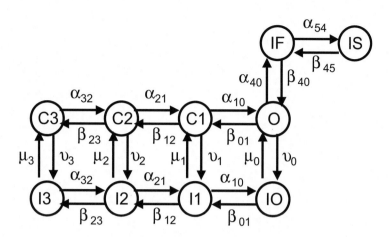

Simulation Problems

(1) Program the Hodgkin-Huxely membrane model and reproduce one of the properties in Sec. 3.3. Assume $G_L = 0.3mS/cm^2$, $g_{na} = 120.0mS/cm^2$, $g_K = 36.0mS/cm^2$, $E_L = -50mV$, $E_{Na} = 55mV$, $E_K = -72.0mV$ $C_m = 1\mu F/cm^2$.

(2) Create a strength-duration curve for anode break of the Hodgkin-Huxely model.

(3) Program the Integrate and Fire membrane model with $R_m = 2k\Omega cm^2$, $C_m = 1.0\mu F/cm^2$, $V_m^{rest} = -70mV$, $V_m^{th} = -60mV$, and $V_m^{peak} = 20mV$. Assume that the voltage is reset to rest after firing.

(4) Find I_{rhe} for the Integrate and Fire model in problem 3. Then, proceed to problem 5.

(5) Program the Hindmarsh-Rose model with the parameters $a = 0.4, b = 2.4, c = 1.5, d = 1.0$, $s = 1.0$, $r = 0.001$. Show that by adding a short-duration I_{stim} to the dx/dt term, these parameters will cause periodic bursting.

(6) For the Hindmarsh-Rose model in problem 5, find a value for a continuous I_{stim} that will turn the bursting into a model that fires periodically.

(7) Program the Traub Neuron model and determine if the properties in Sec. 3.3 apply.

(8) For the Traub model, determine if a continuously applied current can induce repeated bursting. If stimulus strength is changed how does the behavior of the model change?

CHAPTER 4

Propagation

Experimental studies have shown that a neuron does not fire all at once. Rather, there is a wave of electrical activity that is passed from one small patch of membrane to the next, and so on, around the cell. Likewise, electrical impulses spread down dendrites to the soma by moving from one patch of membrane to the next. A similar process is involved in the spread of electrical impulses from the axon hillock down the axon. These waves of electrical activity are called *propagation*. Not all propagation, however, is the same. As the dendrites are composed of passive patches of membrane, they will leak charge out of each patch as propagation moves forward. Therefore, the strength of the impulse will be decreased, or *attenuated*, as an impulse travels from the ends of the dendrite toward the soma. Attenuated propagation is often called *passive propagation*. The axon, on the other hand, is composed of active patches of membrane that can generate currents (e.g., an action potential). Therefore, an impulse that enters the axon hillock will propagate *unattenuated* to the end of the axon. Furthermore, because of the refractory period, once a patch of membrane has fired it can not fire again for some time. The consequence is that propagation can proceed in only one direction. Unattenuated propagation is often called *active propagation*.

In this chapter, we will develop models to describe propagation. As passive propagation is simpler than active propagation, we will begin by considering propagation in the dendrites. Active propagation in axons will be explained as a special case of passive propagation.

4.1 PASSIVE PROPAGATION IN DENDRITES

4.1.1 The Core Conductor Model

The dendrites of a neuron may be modeled as many small cylindrical patches of passive membrane connected together into a thin one-dimensional *cable* (Fig. 4.1). In fact, this assumption is the same as that used by the pioneers of cable theory to describe propagation of electricity down a wire. As the theory of propagation down a wire was developed first, the terms *cable theory* and *core conductor theory* have been adopted by electrophysiologists.

Consider the *discrete* cable in Fig. 4.2 where the passive elements are separated by dx and connected together in the intracellular and extracellular space by a resistance per unit length, r_i and r_e (Ω/cm). From Fig. 4.2, we can choose a node in the center of the cable in Fig. 4.3 and write down intra and extracellular currents at nodes 1, 2, and 3 using Kirchhoff's Current Law.

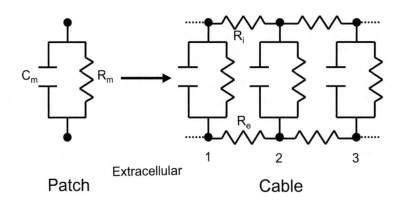

Figure 4.1: Construction of a cable from patches of membrane.

At Node 2,

$$\frac{\phi_i^1 - \phi_i^2}{r_i \cdot dx} - \frac{\phi_i^2 - \phi_i^3}{r_i \cdot dx} - dx \cdot i_m = 0 \tag{4.1}$$

$$\frac{\phi_e^1 - \phi_e^2}{r_e \cdot dx} - \frac{\phi_e^2 - \phi_e^3}{r_e \cdot dx} + dx \cdot i_m = 0 \tag{4.2}$$

or more compactly,

$$\frac{\phi_i^1 - 2\phi_i^2 + \phi_i^3}{dx^2 \cdot r_i} = i_m \tag{4.3}$$

$$\frac{\phi_e^1 - 2\phi_e^2 + \phi_e^3}{dx^2 \cdot r_e} = -i_m \tag{4.4}$$

where i_m at each node is defined by the passive RC circuit:

$$i_m = c_m \frac{dV_m}{dt} - \frac{V_m}{r_m} \tag{4.5}$$

rearranging Eqs. (4.3) and (4.4)

$$\frac{\phi_i^1 - 2\phi_i^2 + \phi_i^3}{dx^2} = r_i \cdot i_m \tag{4.6}$$

$$\frac{\phi_e^1 - 2\phi_e^2 + \phi_e^3}{dx^2} = -r_e \cdot i_m \tag{4.7}$$

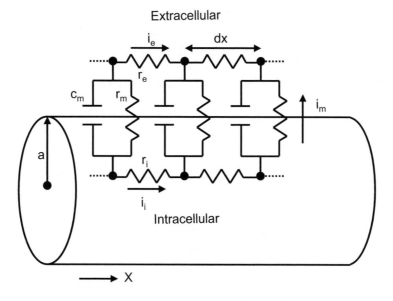

Figure 4.2: Circuit analog of a discrete cable.

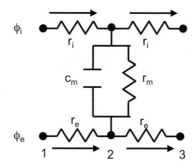

Figure 4.3: Example node in a discrete cable.

and then subtracting Eq. (4.7) from Eq. (4.6)

$$\frac{(\phi_i^1 - \phi_e^1) - 2(\phi_i^2 - \phi_e^2) + (\phi_i^3 - \phi_3^3)}{dx^2} = (r_i + r_e)i_m \tag{4.8}$$

$$\frac{V_m^1 - 2V_m^2 + V_m^3}{dx^2} = (r_i + r_e)i_m \tag{4.9}$$

substitution for i_m yields

$$\frac{V_m^1 - 2V_m^2 + V_m^3}{dx^2} = (r_i + r_e) \left[c_m \frac{dV_m}{dt} + \frac{V_m}{r_m} \right] . \tag{4.10}$$

If we let $dx \to 0$, Eq. (4.10) becomes the continuous cable equation:

$$\frac{\partial^2 V_m}{\partial x^2} = (r_i + r_e) \left[c_m \frac{\partial V_m}{\partial t} + \frac{V_m}{r_m} \right] \tag{4.11}$$

where we have replaced the d/dt terms by $\partial/\partial t$ to indicate that this is a partial differential equation. A bit of algebra yields the following:

$$\frac{1}{r_i + r_e} \frac{\partial^2 V_m}{\partial x^2} = c_m \frac{\partial V_m}{\partial t} + \frac{V_m}{r_m} \tag{4.12}$$

$$\frac{r_m}{r_i + r_e} \frac{\partial^2 V_m}{\partial x^2} = r_m c_m \frac{\partial V_m}{\partial t} + V_m \tag{4.13}$$

or in the typical core conductor form

$$\lambda^2 \frac{\partial^2 V_m}{\partial x^2} = \tau_m \frac{\partial V_m}{\partial t} + V_m \tag{4.14}$$

where $\tau_m = r_m c_m$ is the *membrane time constant* in *msec*, as defined in Sec. 2.2 and

$$\lambda = \sqrt{\frac{r_m}{r_i + r_e}} \tag{4.15}$$

is the *cable space constant* in units of *cm*.

4.1.2   A Simplification
One very common simplification to the cable equation is achieved by assuming that the extracellular bath is much more conductive than the intracellular space. If one assumes $r_e \approx 0$, all of the extracellular potentials are equal and act as an electrical ground. Therefore, $V_m = \phi_i$ and $\lambda = \sqrt{\frac{r_m}{r_i}}$.

4.1.3   Units and Relationships
The units of the variables in Eqs. (4.1)–(4.15) can be confusing because they are different than the units used in Ch. 2 and 3. The tables show the units used in parameters of the core conductor model and their relationship to the membrane parameters.
Given these relationships we can rewrite Eq. (4.14) as:

Table 4.1:		
Parameter	Name	Unit
r_m	Membrane Resistance	Ωcm
r_i	Axial Intracellular Resistance	Ω/cm
r_e	Axial Extracellular Resistance	Ω/cm
c_m	Axial Membrane Capacitance	$\mu F/cm$
i_m	Axial Membrane Current	$\mu A/cm$
dx	Spatial Step Size	cm
a	Cable radius	cm

Table 4.2:			
Variable	Name	Unit	Equivalence
R_i	Specific Intracellular Resistivity	Ωcm	$\pi a^2 r_i$
R_e	Specific Extracellular Resistivity	Ωcm	$\pi a^2 r_e$
R_m	Specific Membrane Resistivity	Ωcm^2	$2\pi a r_m$
C_m	Specific Membrane Capacitance	$\mu F/cm^2$	$c_m/(2\pi a)$
I_m	Specific Membrane Current	$\mu A/cm^2$	$i_m/(2\pi a)$

$$\frac{1}{(R_i + R_e)} \frac{\partial^2 V_m}{\partial x^2} = \beta \left[C_m \frac{\partial V_m}{\partial t} + \frac{V_m}{R_m} \right] \tag{4.16}$$

where $\beta = 2/a$.

4.1.4   An Applied Stimulus

A stimulus, i_{stim}, may be applied at a particular location on on an infinitely long cable for some particular duration. Mathematically, a stimulus applied to a point can be represented using the Dirac delta function, $\delta(x)$. If the stimulus at this point is applied at $t = 0$ and remains on, the unit step function, $u(t)$, may be used. For the simple case where a stimulus is applied to a point in the middle of an infinitely long cable beginning at $t = 0$,

$$\lambda^2 \frac{\partial^2 V_m}{\partial x^2} = \tau_m \frac{\partial V_m}{\partial t} + V_m \pm r_m i_{\text{stim}} \delta(x) u(t) \tag{4.17}$$

where the $\pm$ is to take into account either an intracellular or extracellular stimulus.

4.1.5   Steady-State Solution

We can consider the steady state solution to Eq. (4.17) by assuming $\frac{\partial V_m}{\partial t} = 0$:

$$\lambda^2 \frac{d^2 V_m}{dx^2} - V_m = -r_m i_{\text{stim}} \delta(x) \tag{4.18}$$

where we have assumed the stimulus will depolarize the membrane. Equation (4.18) is an ordinary differential equation that no longer depends upon time. We can find the homogeneous solution by finding the solution to:

$$\lambda^2 \frac{d^2 V_m}{dx^2} - V_m = 0 . \tag{4.19}$$

The solution to Eq. (4.19) is the Helmholtz equation:

$$V_m(x) = Ae^{-|x|/\lambda} + Be^{|x|/\lambda} . \tag{4.20}$$

The second term does not make physical sense so

$$V_m(x) = Ae^{-|x|/\lambda} . \tag{4.21}$$

To find A we will integrate Eq. (4.18) around the stimulus ($x = 0$):

$$\lambda^2 \int_{0-}^{0+} \frac{d^2 V_m}{dx^2} \, dx - \int_{0-}^{0+} V_m \, dx = -r_m i_{\text{stim}} \int_{0-}^{0+} \delta(x) \, dx . \tag{4.22}$$

Evaluation shows that $A = \frac{r_m \cdot i_{\text{stim}}}{2\lambda}$ so the steady-state solution is

$$V_m(x) = \frac{r_m i_{\text{stim}}}{2\lambda} e^{-|x|/\lambda} . \tag{4.23}$$

Note that the value of the constant $\frac{r_m i_{\text{stim}}}{2\lambda}$ will be dependent upon the nature of the stimulus. For example, we have assumed that the extracellular space is large and very conductive.

4.1.6   Finding the Length Constant

If a cable with unknown membrane properties is encountered, Eq. (4.23) is a way to find λ and r_m. If a known stimulus is applied for a long time, the membrane will eventually reach steady-state. At the point of the stimulus, $x = 0$, so the exponential term becomes 1. Therefore, the voltage at the stimulus is $\frac{r_m \cdot i_{\text{stim}}}{2\lambda}$. Furthermore, this voltage level will fall off in space (in both directions because of $|x|$) at an exponential rate governed by λ. Therefore, λ can be found as the rate of fall off in a similar way to finding τ_m in Ch. 2. Once λ is known, r_m can be found from $\frac{r_m \cdot i_{\text{stim}}}{2\lambda}$.

4.1.7   Time and Space Dependent Solution

Although we will not show the solution here, it is possible to solve Eq. (4.17) for V_m as a function of both time and space:

$$V_m(x, t) = \frac{r_m \cdot i_{\text{stim}}}{4\lambda} \left[e^{-|x|/\lambda} erfc\left(\frac{|x|}{2\lambda}\sqrt{\frac{\tau_m}{t}} - \sqrt{\frac{t}{\tau_m}} \right) \right.$$
$$\left. - e^{|x|/\lambda} erfc\left(\frac{|x|}{2\lambda}\sqrt{\frac{\tau_m}{t}} + \sqrt{\frac{t}{\tau_m}} \right) \right] \tag{4.24}$$

where *erfc* is the complimentary error function defined by $erfc(y) = 1 - erf(y)$ and

$$erf(y) = \frac{2}{\pi} \int_y^0 e^{-z^2} \, dz \tag{4.25}$$

is defined as the error function (see Fig. 4.4). It is important to note that Eq. (4.24) is for an infinite cable with a negative i_{stim}. The leading terms in Eqs. (4.23) and (4.24) will change if the nature of the stimulus is changed or the cable is not infinitely long.

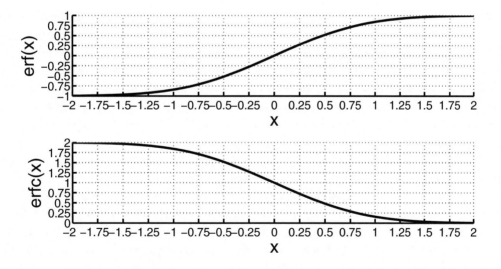

Figure 4.4: Error and complimentary error functions.

Figure 4.5 shows the time and space solutions to Eq. (4.24) for different times (left panel) and different locations in space (right panel). Because the specific numbers may vary for any given cable, the plot is shown in terms of the general passive properties of the cable, i.e., λ and τ_m. It is important to note that the fall off in space is exponential only for the steady-state ($t \to \infty$). Likewise, the rise in time is exponential (as in an RC circuit) only at $x = \lambda$.

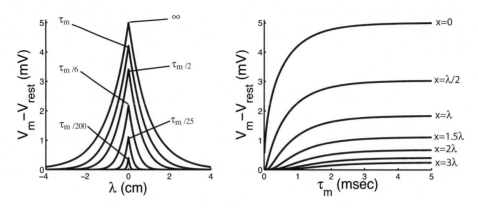

Figure 4.5: Time and space solution in a passive cable.

4.2   ACTIVE PROPAGATION IN AXONS

In the previous section, we considered passive propagation down a cable as a model of dendritic propagation. The membrane of the axon, however, has many nonlinear ion channels that are capable of generating an action potential. The only modification needed to create an *active cable* is to replace the passive $I_{ion} = V_m/r_m$ with a more complex model as outlined in Ch. 3. Whereas a stimulus applied to a passive cable will be *attenuated*, in an active cable the action potential will propagate unattenuated. In this way, a signal that reaches the axon hillock will be propagated unattenuated to the end of the axon. In all but the simplest cases, it is not possible to derive an analytic solution when propagation is active.

To demonstrate the concept of active propagation, Fig. 4.6 shows propagation down an active $3cm$ long axon. The superthreshold stimulus was delivered to the left end of the axon at $t = 0msec$. In the left panel, action potentials are shown at $x = 1cm$ and $x = 2cm$. Notice that although the action potential shape does not change, there is a delay in timing of the upstroke. The right panel shows a snapshot of the voltage along the cable at $t = 40msec$. Notice that when propagation is from left to right, the spatial plot has the shape of a reversed action potential. To understand why, consider the location at $x = 2cm$. In the right panel, at $40msec$, the point is at rest. The shape in the left panel, however, is moving the right so eventually the sharp spike will reach $x = 2cm$. From the right panel we know that this occurs at approximately $t = 55msec$. In space, the reversed action potential shape will continue to move the right, causing the location at $x = 2cm$ to undergo all of the phases of an action potential.

4.2.1   Uniform Propagation
If I_{ion} and the cable properties are constant everywhere in the cable, we can derive that

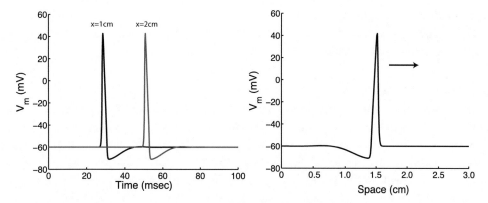

Figure 4.6: Propagation down an active cable.

$$V_m(x, t) = V_m[x - \theta(t)] \tag{4.26}$$

where θ is the *propagation velocity* is a measure of the speed at which the action potential moves down the cable. The reason Eq. (4.26) works is because the shape of the action potential is the same at every point on the cable. The only parameter that changes is *when* the action potential occurs in time. Using the chain rule twice on Eq. 4.26

$$\frac{\partial V_m}{\partial x} = \frac{1}{\theta} \frac{\partial V_m}{\partial t} \tag{4.27}$$

$$\frac{\partial^2 V_m}{\partial x^2} = \frac{1}{\theta^2} \frac{\partial^2 V_m}{\partial t^2}. \tag{4.28}$$

Substitution into Eq. (4.16) yields

$$I_m = \frac{a}{2R_i\theta^2} \frac{d^2 V_m}{dt^2}. \tag{4.29}$$

Upon careful inspection, Eq. (4.29) has a deeper meaning. As long as I_{ion} does not change throughout the cable, a family of solution exists with the only requirement being:

$$\frac{a}{2R_i\theta^2} = K \tag{4.30}$$

$$\theta = \sqrt{\frac{a}{2R_i K}} \tag{4.31}$$

where K is a constant and dependent only on membrane properties. The usefulness of Eqs. (4.30) and (4.31) is that only one set of a and θ is needed to find $R_i K$. Once $R_i K$ is known, the effect of any change in a on θ can be predicted.

4.2.2   Saltatory Conduction

A problem with using active propagation to transmit an impulse over long distances is that the signal will take time to traverse the axon. In fact, in some regions of the body an axon can be up to one meter long. To compensate, the nervous system has developed a clever solution to speed up propagation. Schwann cells (a type of glial cell) create a *myelin sheath* around the axon. The presence of this insulating sheath makes it nearly impossible for current to cross the cell membrane. The impact is that membrane directly under the sheath has a much higher R_m. The thicker membrane also decreases C_m. Therefore, membrane covered in myelin is effectively *passive*.

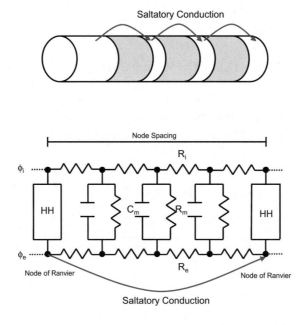

Figure 4.7: Schematic of saltatory conduction.

As shown in Fig. 4.7, the sheath does leave small regions of the neuron cell membrane exposed, called *Nodes of Ranvier*, which may fire an action potential. If an active patch of membrane fires at a Node of Ranvier, current will flow to the right but will not be able to easily cross the membrane because of the high R_m. Instead, most current will follow the path of least resistance and jump farther down the cable. So, in effect, the high resistance of the myelin creates a short circuit that skips quickly from one Node of Ranvier to the next. Multiple Sclerosis (MS) is a disease that causes inflammation and scaring of the myelin sheath leading to degradation of neural impulse propagation. The symptoms are changes in sensations, muscle spasms, a lack of coordination and balance, pain, and eventually cognitive and emotional impairment.

4.3   PASS THE PAPER PLEASE

Propagation occurs in many other systems, using chemical, mechanical, fluid, or bits of information as the medium of propagation. For an alternative perspective on propagation, consider arranging you and ten of your classmates in a line all shoulder to shoulder. To the right of the line is a large bag filled with shredded paper. The person on the far right end then grabs a large handful of shreds in their right-hand. The rules for each person will be to receive paper in their right-hand, pass it to their left-hand and then pass to the right-hand of the next person in the line. Using these simple rules, the shreds will *propagate* down the line from right to left. If your line tries to propagate shreds quickly, you will notice that most of the shreds are lost by the time propagation ends at the left end of the line. In the activity, the shreds are analogous to charges and the movement of shreds is a current flow. The size of each person's hands are analogous to a capacitance and how much paper they lose is similar to a leakage resistance. You may also notice that much of the paper has fallen to the ground. In fact, this is the same situation as when charge leaves the intracellular space of the neuron to disappear into the large extracellular space. Next, consider that most of the shreds are lost in the first few transitions with much less being lost as propagation continues. Therefore, the way shreds are lost as propagation continues down the cable may be approximated by a decaying exponential.

As the shreds are lost during propagation, our analogy is to passive propagation. A small change, however, could turn passive propagation to active propagation. If *everyone* in the line has a handful of shreds in their left hand, and the propagation is reinitiated, you will discover that the person at the end of the line will have a considerable amount of shreds. Furthermore, the active propagation activity could be modified for saltatory conduction by simply skipping every other (or every three) people in the line.

4.4   NUMERICAL METHODS: THE FINITE AND DISCRETE CABLE

In all but the most simple situations, the continuous cable model can only be solved numerically using a computer. It is therefore required that the cable be of finite length and divided into small patches of membrane. Beginning with the continuous Eq. (4.11) the *finite difference* approximation may be made for the second derivative in space.

$$\frac{V_m^{k-1} - 2V_m^k + V_m^{k+1}}{(r_i + r_e)\,dx^2} = \left[c_m \frac{dV_m}{dt} + \frac{V_m}{r_m} \right] \tag{4.32}$$

where k is a variable to represent any node in the middle of the cable. Note that Eq. 4.32 is a general form of Eq. (4.10). Rearranging

$$\frac{dV_m}{dt} = \frac{1}{c_m}\left[\frac{V_m^{k-1} - 2V_m^k + V_m^{K+1}}{(r_i + r_e)\,dx^2} - \frac{V_m}{r_m}\right] \tag{4.33}$$

and using the Euler method of Sec. 2.6

$$dV_m = \frac{dt}{c_m}\left[\frac{V_m^{k-1} - 2V_m^k + V_m^{k+1}}{(r_i + r_e)\,dx^2} - \frac{V_m}{r_m}\right] \tag{4.34}$$

$$V_m^{t+\Delta t} = V_m^t + dV_m \tag{4.35}$$

and every V_m is discrete in both time and space. Note that in active propagation the gating variable must also be integrated in time, however, they do not require information from their neighbors. Therefore, the simple Euler method may be used.

A problem is encountered, however, with the above formulation. Consider evaluating the left most node on the cable, e.g., $k = 1$. An update of the end node requires information at $k = 0$ which does not exist. The same problem occurs at the right most node of the cable, e.g. $k = n$. To overcome the problem at the endpoints, we must enforce a boundary condition. The most common assumption in neural propagation is that no current can leave either end of the cable. This is called a *sealed-end* boundary condition. Mathematically,

$$\frac{d\phi_i}{dx} = -I_i r_i = 0 \tag{4.36}$$

$$\frac{d\phi_e}{dx} = -I_e r_e = 0 . \tag{4.37}$$

The most straightforward way to enforce this boundary condition is to define a *ghost* node that extends past the end of the cable (see Fig. 4.8). Then, at the two cable ends

$$\phi_{i,e,}^{gleft} = \phi_{i,e,}^2 \tag{4.38}$$

$$\phi_{i,e,}^{gright} = \phi_{i,e}^{n-1} \tag{4.39}$$

and ensures that at the first and last node, Eqs. (4.36) and (4.37) are satisfied.

Combining Eq. (4.34) and the sealed-end, we can write the equation for node 1 as

$$dV_m = \frac{dt}{c_m}\left[\frac{V_m^{gleft} - 2V_m^1 + V_m^2}{(r_i - r_e)\,dx^2} - \frac{V_m}{r_m}\right] \tag{4.40}$$

$$dV_m = \frac{dt}{c_m}\left[\frac{2V_m^2 - 2V_m^1}{(r_i - r_e)\,dx^2} - \frac{V_m}{r_m}\right] . \tag{4.41}$$

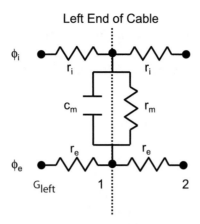

Figure 4.8: Sealed-end boundary condition.

Other possible boundary conditions for the end of the cable are to allow current to leave (*leaky end*) or to clamp the voltage (*clamped end*).

4.5   NUMERICAL METHODS: TEMPLATE FOR CABLE PROPAGATION

In Sec. 2.6, a way of numerically solving a differential equation was outlined. In the active membrane models, we need to keep track of several differential equations as well as compute rate constants, steady-state values and currents. Below is a template for how to write a program to solve the active equations.

```
Define constants (e.g., R_i, R_e, a, C_m, other membrane variables)
Compute initial αs and βs
Compute initial conditions for state variables
(e.g., V_m^rest, m, h, n)

for (time=0 to time=end in increments of dt)
    for (i=1 to i=Last Node in increments of 1

        Compute Currents at node i
        Compute αs and βs at node i

        if (i=1)
            Update Differential Equations at left boundary
```

```
        else if (i=last node)
            Update Differential Equations at right boundary
        else
            Update Differential Equations at all middle nodes
        end

        Save values of interest into an array (e.g., Vm(i))

    end
end

    Store values of interest to a file
```

Summary

(1) Patches of neuron membranes can share current with neighbors. The result is propagation of an electrical impulse.

(2) The core conductor model is a circuit analogy of how electrical impulses are shared between neighboring patches of membrane. It requires consideration of both time and space.

(3) The key membrane properties in determining how electrical impulses propagate are the membrane time constant and the membrane space constant. These properties may be found experimentally by applying a controlled stimulus and measuring the decay rates of transmembrane voltages in time and space.

(4) Passive propagation results when each patch of membrane is not capable of generating current.

(5) Active propagation results when each patch of membrane is capable of generating current.

Homework Problems

(1) Work out the units for Eqs. (4.13)–(4.16).

(2) Show the steps between Eq. (4.22) and Eq. (4.23).

(3) Show how Eq. (4.24) becomes Eq. (4.23) as $t \to \infty$.

(4) Using Eq. (4.24), show that at $x = \lambda$, the time constant, τ_m, can be found as 63% of the steady-state value. Show that at $x = 0$, τ_m can be found as 84% of the steady-state value.

(5) A passive nerve of radius $50\mu m$ is stimulated with a current pulse of $10nA$ in the middle $(x = 0.0cm)$ of and cable. Using the plots below to answer the questions:

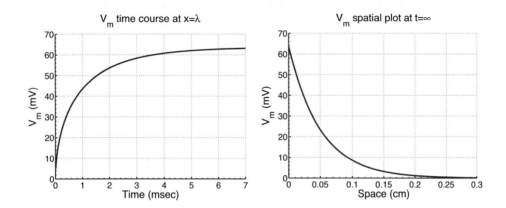

a) Find τ in $msec$.
b) Find λ in cm. Start with the full solution to the cable equation.
c) What is the R_m in Ωcm^2?

(6) In Fig. 4.6:
a) What is the propagation velocity in cm/s?
b) Explain what would happen to the spatial distribution if the propagation velocity was slower.
c) Explain what would happen to the spatial distribution if the propagation direction was reversed?
d) Explain what would happen to the spatial distribution if the propagation was saltatory.

(7) In Fig. 4.6, assume $a = 10\mu m$, and predict the propagation velocity if a is changed to $a = 13\mu m$.

(8) In Eq. 4.18, use the definition of membrane current to determine when the stimulus is positive and when it is negative with regard to an intracellular or extracellular current carried by a negative ion.

Simulation Problems

(1) Write code to simulation a passive $3cm$ cable with $R_m = 1k\Omega cm^2$, $C_m = 1\mu F/cm^2$, $dx = 100\mu m$, $a = 10\mu m$, $R_i = 100\Omega cm$ and assume that the extracellular space is large and acts as

a ground and the boundaries are sealed-ends. Demonstrate that your problem is working by reproducing the left and right panels of Fig. 4.5.

(2) Modify the program from 1 above for active propagation of a Hodgkin-Huxely action potential. Stimulate the left end of the cable with a strength and duration sufficient to begin propagation. Demonstrate that your program works by creating a figure similar to 4.6.

(3) Use the code from 2 above to demonstrate the impact of changing the cable radius on propagation velocity.

(4) Use the code from 2 above to demonstrate the impact of changing R_i.

(5) Modify the program from 2 to demonstrate how saltatory conduction speeds propagation velocity. Let $R_m = 10k\Omega$ and $C_m = 0.01\mu F/cm^2$ for the sheath.

CHAPTER 5

Neural Branches

In Ch. 4, we considered propagation in uniform and straight cables. Figure 5.1 shows the complex web of dendrites that is collectively called the *dendritic tree*. It is clear that the tree contains branching cables, changes in diameter and curves. In this chapter we will generalize cable theory to handle changes in membrane properties and branching. We will first examine a method of simplifying the tree by *lumping* the dendrites into one large compartment. We will then turn to the more sophisticated method of *multicompartment* models.

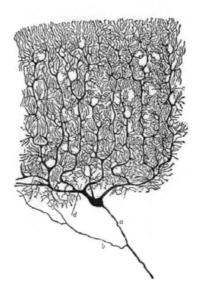

Figure 5.1: Dendritic tree.

5.1 LUMPED MODELS

Lumped models were developed before computers became available as a way to simplify a complex neuron to a single unit. The goal was to make analytical predictions about how an impulse at the end of a dendrite would be attenuate on its way toward the soma.

5.1.1   The Rall Model

To model the complex dendritic tree, Rall recognized that the branching dendrite was simply acting as a receiver for many inputs and then attenuating the input. He therefore proposed that the dendritic tree could be collapsed down to one signal passive cable of varying diameter. This would mean that the dendritic tree could be represented by an *equivalent* cable as shown in Fig. 5.2. Likewise, the various stimuli to the ends of the dendrites would be approximated by a single stimulus. The advantage of this idea is that computation would become fast and efficient and the actual geometry of the tree would not matter. It also enabled the analytical solution of attenuation down the dendritic tree. The disadvantage was that to achieve this collapse of the tree, Rall had to assume a relationship between the radii of all branches in relation to their main trunk

$$a_{main}^{3/2} = a_{branch1}^{3/2} + a_{branch2}^{3/2} + \dots . \tag{5.1}$$

This relationship was assumed to apply to *every* branch in the tree. Although Eq. (5.1) appears to be very restrictive, Rall showed that for many dendritic trees is was a reasonable approximation.

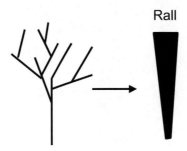

Figure 5.2: Rall's equivalent cable.

5.1.2   The Ball and Stick Model

Another advantage of the collapsed dendritic tree is that the equivalent cable can be attached to a soma and axon to create a *ball and stick* model of a neuron. In Fig. 5.3, the dendrites have been collapsed using the method outlined by Rall and the soma is modeled as a single compartment which may be either active or passive. The axon is in fact not present at all and is simulated by a time delay. As we found in Ch. 4, once an action potential has fired in the soma, it will propagate down the axon unattenuated to the axon terminal. So, the delay depends only upon the speed of propagation and length of the axon. All of these simplifications allow for the behavior of a neuron to be solved numerically with a minimum of computing power.

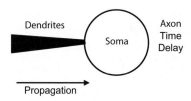

Figure 5.3: Ball and stick model.

5.2   MULTICOMPARTMENT MODELS

It is clear from Fig. 5.1 that not all neurons will follow the assumption of Eq. (5.1). To lift the assumptions of the lumped models, we can consider that the dendrites and axons are made up of many small compartments. The general concept is shown in Fig. 5.4 and is a similar idea to when we derived the original cable equation using many small patches of membrane. Again, current is passed from one compartment to another in one direction only. At a branch, the current is simply split and passed into two (or more) compartments. Below we will derive equations to describe how this current split is achieved. *Multicompartment models* not only remove the Rall restriction but also allow for many different types of post-synapses to be incorporated into the model. The post-synapse will be considered in more detail in Ch. 6. The disadvantage of the multicompartment models is that compared to the Rall model, they must be solved numerically and required considerably greater computing power. Below we will introduce the basic elements of how to construct a dendritic tree of any complexity.

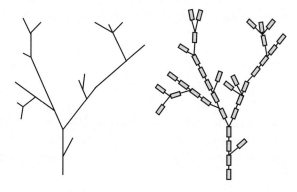

Figure 5.4: Schematic of dendritic compartments.

5.2.1   A Simple Compartment

To begin, we will make a slight change to the definition of a compartment by placing a node at the *center* of each patch as in Fig. 5.5. Each compartment will therefore have a length of $2dx$ and each half resistor in a compartment will be defined as

$$r_i = R = \frac{dx \cdot R_i}{\pi a^2} .$$ (5.2)

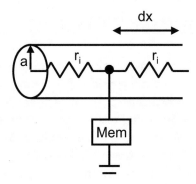

Figure 5.5: Model of a single compartment.

5.2.2   Change in Fiber Radius

Since the composition of the cytoplasm does not change drastically, the intracellular resistivity (r_i, a material property) does not change. Therefore, the determining factor of the half resistance with a radius, a, is

$$R(a) = \frac{dx \cdot R_i}{\pi a^2} .$$ (5.3)

Using the function $R(a)$, we can consider the impact of coupling a patch with a small radius (a) to a patch with a larger radius (b) as in Fig. 5.6. If we assume that all current flows from left to right and that extracellular space is conductive (e.g., $r_e = 0$, $V_m = \phi_i$), then at Node 1

$$\frac{V_m^0 - V_m^1}{[R(a) + R(a)]} - \frac{V_m^1 - V_m^2}{[R(a) + R(b)]} - I_m = 0$$ (5.4)

and at Node 2

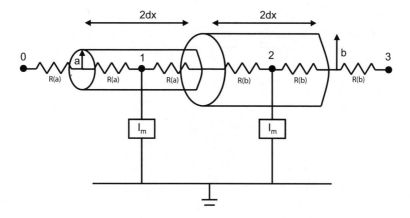

Figure 5.6: Impact of a radius change.

$$\frac{V_m^1 - V_m^2}{[R(a) + R(b)]} - \frac{V_m^2 - V_m^3}{[R(b) + R(b)]} - I_m = 0 \, . \tag{5.5}$$

Although these equations are not compact, they are easy to compute numerically. Because R is also dependent upon dx, it is simple to extend our result to compartments of different lengths.

5.2.3 Branches

It is clear from Fig. 5.1 that we also must describe the points at which the dendrite branches. For simplicity, we will consider a single branch, as shown in Fig. 5.7, and assume that dx and a remain constant. The node labeled b is known as a *branch* node. There is no membrane at the branch node because it is not at the center of a compartment and is only used to make computation easier. At Node b,

$$\frac{V_m^1 - V_m^b}{R(a)} - \frac{V_m^b - V_m^2}{R(b)} - \frac{V_m^b - V_m^3}{R(b)} = 0 \, . \tag{5.6}$$

If $a = b$, then $R(a) = R(b)$ and

$$V_m^b = \frac{V_m^1 + V_m^2 + V_m^3}{3} \, . \tag{5.7}$$

Note that if dx or a were not the same everywhere, Eq. 5.7 would become much more complicated but still computable. We can now use V_m^b to derive equations for Nodes 1, 2, and 3. At Node 1

$$\frac{V_m^0 - V_m^1}{2R} - \frac{V_m^1 - V_m^b}{R} - I_m^1 = 0 \tag{5.8}$$

Similarly:

$$\frac{V_m^b - V_m^2}{R} - \frac{V_m^2 - V_m^4}{2R} - I_m^2 = 0 \tag{5.9}$$

$$\frac{V_m^b - V_m^3}{R} - \frac{V_m^3 - V_m^5}{2R} - I_m^3 = 0 \tag{5.10}$$

$$\tag{5.11}$$

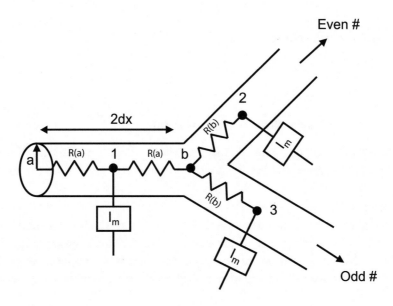

Figure 5.7: Circuit analog of a neural branch.

The remainder of the extension of these branches (e.g., Nodes 4, 5, 6, ...) can be formulated as uniform cables. Therefore, the key to writing the equations for a branch is the branch node.

5.2.4 The Soma

Although the Soma is considered to be the center of the neuron, it is often modeled as a single large passive compartment. The special role of the soma, however, is to integrate all impulses entering the dendritic tree. It is also directly connected to the axon hillock which *is* capable of generating an action potential. So, if the dendrites charge up the soma above threshold, the axon hillock will fire and send an action potential propagating down the axon. The soma therefore acts as a relay center between the dendrites and axon. Figure 5.8 shows a simple model of a neuron composed of two active axon compartments of radius a, a passive soma of radius b, a main dendritic trunk of radius c, two branches of radius d, and a third branch of radius e.

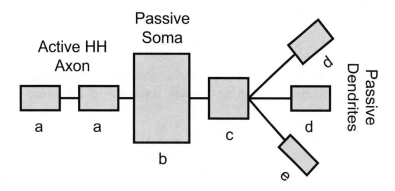

Figure 5.8: A structural model of a simple neuron.

By combining the ideas of active and passive cables with the ability to create branches and changes in diameter, we have all the concepts needed to create compartment models of any complexity. In fact, some researchers have created neuronal models, based upon the histology of real neurons that contain hundreds (or even thousands) of compartments. Although most of these models are two-dimensional, the technology exists to create three dimensional images of the branching structures of real dendrites. You may wish to think about how the analysis above needs to be changed to allow for a 3D structure.

5.2.5   Axon Collaterals and Semi-Active Dendrites

To simplify the discussion above we have made a number of assumptions that may not be true in a real neuron. First, some specialized neurons have either multiple axon projections or a single axon that branches many times. The geometry of a branching axon can be modeled in the same way as a branching dendrite with the exception that the ionic current, I_{ion} will be more complex.

Recent data has also shown that dendrites are not simply passive cables. In other words, there are some ion channels embedded in the membrane that have nonlinear behavior so $I_m \neq \frac{V_m}{R_m}$. To model a *semi-active* dendrite we can simply replace the passive I_{ion} term with the appropriate nonlinear currents. The presence of these channels allows currents and potentials to *back propagate* in the orthodromic direction. Back propagation may impact future pulses in the antidromic direction by changing the concentrations of ions (in particular $[Ca^{2+}]_i$) and transiently change R_m of the dendrite.

Lastly, anatomical studies have shown that small protrusions, called *spines*, appear on some dendrites. Although it is not clear what role spines play, there have been a number of suggestions. One possibility is that spines increase the amount of membrane area for synaptic inputs from other neurons. Another view is that spines store $[Ca^{2+}]_i$ and other ions which modulate the strength of synaptic inputs. Lastly, some consider the spines to serve an electrical role by modulating the attenuation of a voltage headed toward the soma by acting as small current amplifiers.

5.3   NUMERICAL METHODS: MATRIX FORMULATION

A typical node, k, in an unbranched cable of length n with uniform membrane properties is connected to two neighbors. The governing equation for this node is therefore of the form:

$$\frac{V(k-1) - 2V(k) + V(k+1)}{R} = I_m(k) . \tag{5.12}$$

At the ends of the cable we can assume sealed ends as in Sec. 4.4

$$\frac{-2V(1) + 2V(2)}{R} = I_m(1) \tag{5.13}$$

$$\frac{2V(n-1) - 2V(n)}{R} = I_m(n) . \tag{5.14}$$

Using matrices we can write this system of equations as

$$
\begin{bmatrix}
\frac{-2}{R} & \frac{2}{R} & \cdots & & & & \\
\frac{1}{R} & \frac{-2}{R} & \frac{1}{R} & \cdots & & & \\
0 & \frac{1}{R} & \frac{-2}{R} & \frac{1}{R} & \cdots & & \\
\vdots & \vdots & \vdots & & \ddots & & \\
& & & \cdots & \frac{1}{R} & \frac{-2}{R} & \frac{1}{R} & 0 \\
& & & \cdots & & \frac{1}{R} & \frac{-2}{R} & \frac{1}{R} \\
& & & \cdots & & & \frac{2}{R} & \frac{-2}{R}
\end{bmatrix}
\begin{bmatrix}
V(1) \\
V(2) \\
V(3) \\
\vdots \\
V(n-2) \\
V(n-1) \\
V(n)
\end{bmatrix}
=
\begin{bmatrix}
I_m(1) \\
I_m(2) \\
I_m(3) \\
\vdots \\
I_m(n-2) \\
I_m(n-1) \\
I_m(n)
\end{bmatrix}
$$

or including a more detailed representation for I_m

$$
\begin{bmatrix}
\frac{-2}{R} & \frac{2}{R} & \cdots & & & & \\
\frac{1}{R} & \frac{-2}{R} & \frac{1}{R} & \cdots & & & \\
0 & \frac{1}{R} & \frac{-2}{R} & \frac{1}{R} & \cdots & & \\
\vdots & \vdots & \vdots & & \ddots & & \\
& & & \cdots & \frac{1}{R} & \frac{-2}{R} & \frac{1}{R} & 0 \\
& & & \cdots & & \frac{1}{R} & \frac{-2}{R} & \frac{1}{R} \\
& & & \cdots & & & \frac{2}{R} & \frac{-2}{R}
\end{bmatrix}
\begin{bmatrix}
V(1) \\
V(2) \\
V(3) \\
\vdots \\
V(n-2) \\
V(n-1) \\
V(n)
\end{bmatrix}
$$

$$= C_m \frac{d}{dt} \begin{bmatrix} V(1) \\ V(2) \\ V(3) \\ \vdots \\ V(n-2) \\ V(n-1) \\ V(n) \end{bmatrix} + \begin{bmatrix} I_{\text{ion}}(1) \\ I_{\text{ion}}(2) \\ I_{\text{ion}}(3) \\ \vdots \\ I_{\text{ion}}(n-2) \\ I_{\text{ion}}(n-1) \\ I_{\text{ion}}(n) \end{bmatrix}$$

or more compactly,

$$\mathbf{AV} = C_m \frac{d\mathbf{V}}{dt} + \mathbf{I_{ion}} \tag{5.15}$$

where the bold text indicates a vector or matrix. $\mathbf{A}$ is called a *coupling* matrix because the location of the entries show exactly how nodes are connected. Each row of $\mathbf{A}$ corresponds to one node. Although we have assumed R is constant, in reality we know that the resistance will depend upon a and dx. In principle, R could vary in the matrix $\mathbf{A}$ and will indicate the strength of the connections.

To solve Eq. (5.15) we can perform the same rearrangement as in Sec. 2.6

$$d\mathbf{V} = \frac{dt}{C_m} \left[\mathbf{AV} - \mathbf{I_{ion}} \right] \tag{5.16}$$
$$\mathbf{V_{new}} = \mathbf{V_{old}} + d\mathbf{V} . \tag{5.17}$$

Besides being a compact way of writing the simultaneous equations of a cable, there is a practical reason for the *vectorized* form of Eq. (5.16). There exist many sophisticated methods of multiplying, adding, and factoring vectors and matrices that can drastically speedup the simulations.

Summary

(1) The dendrites of a neuron form a branching cable with varying diameter.

(2) The branching cables may be treated as one larger cable (a lumped model) if certain conditions are met.

(3) If these certain conditions are not met, core conductor theory can still be used to include branches and changes in radii.

(4) The derivation of branching cables applies to both passive and active membranes.

Homework Problems

(1) Create the **A** coupling matrix for a cable where $R_i = 100\Omega cm$.

(2) How does the form of the matrix derived in problem 1 change if the cable branches?

(3) Create the coupling matrix for the simple neuron in Fig. 5.8. Assume sealed end boundaries.

(4) Explain how to create the matrix for the right side of Fig. 5.4 assuming all of the radii are equal.

(5) The Traub model of a Pyramidal neuron is shown in Fig. 7.1. Create the coupling matrix for this neuron assuming the diameter of all compartments is the same.

Simulation Problems

(1) Write a computer program to simulate the simple neuron in Fig. 5.8 where the passive properties are as given in Ch. 4, simulation problem 1, and the active properties are determined by HH parameters. Demonstrate that by stimulating one branch with a large enough current that an action potential is fired in the soma and propagates down the axon.

(2) In the program created above, does the stimulus threshold change if you stimulate the branch with radius e versus stimulating one of the branches with radius d.

(3) There are two primary ways that the soma could reach threshold. First, a single large current may depolarize one branch, which although attenuated, may still be above the threshold of the soma. Second, many smaller currents may enter many branches simultaneously and be *integrated* (added up) to bring the soma to threshold. Use the program in 1 to demonstrate both ways of firing the soma.

CHAPTER 6

Synapses

Previous chapters have described how electrical activity propagates down the dendrites, integrates at the soma, and propagates down the dendrites. Most of the action, however, occurs at the connections between neurons. There are two primary ways that neurons connect. First are *gap junctions* which are proteins that directly connect the intracellular space of two neurons. Typically, gap junctions are modeled as a linear resistor.

In this chapter, we will focus primarily on the second type of connection, the synapse. Using engineering terms, the electrical signal that reaches the axon terminal is *transduced* to a chemical signal by the pre-synapse. This chemical signal then diffuses across the synaptic cleft and is transduced back to an electrical signal by the post-synapse on the dendrite of another neuron. A molecule that can transduce an electric signal across a synapse is called a *neurotransmitter*. In reality, these three steps are much more complex and have been broken down in the list below and shown graphically in Fig. 6.1.

1. Neurotransmitters are packaged into vesicles in the soma.

2. Depolarizations in the dendrites charge the soma and cause the axon hillock to fire an action potential.

3. The action potential propagates unattenuated down the axon.

4. The action potential reaches the pre-synaptic terminal.

5. Ca^{2+} channels in the pre-synapse open and Ca^{2+} enters the pre-synapse.

6. Ca^{2+} in the pre-synapse triggers fusion of vesicles (filled with neurotransmitter molecules) with the membrane.

7. The neurotransmitter is released into the synaptic cleft by exocytosis.

8. The neurotransmitter diffuses across the synaptic cleft.

9. The neurotransmitter binds to receptors on the post synaptic membrane.

10. Post-synaptic ion channels open by either a direct or indirect (second messenger) mechanism.

11. The post-synaptic membrane potential changes (either depolarizes or hyperpolarizes).

12. The neurotransmitter is released by the post-synapse, inactivated and finally returned to the pre-synapse.

We have already quantified steps 2–4 and we will not quantify 1 or 7. In this chapter, we will first consider models of the pre-synapse (steps 5 and 6) and then turn to models of the post-synapse (steps 8–12). Although it is typical to examine synapses that connect an axon of one neuron to the dendrite of another, synapses also may form directly on the soma. First, however, the action of neurotransmitters will be discussed.

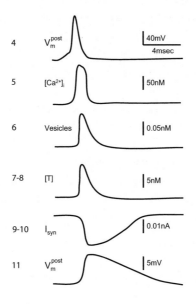

Figure 6.1: Voltages and concentrations of the synapse.

6.1   NEUROTRANSMITTERS

Neurotransmitters are the chemicals that relay electrical signals from the pre-synapse of one neuron to the post-synapse of another neuron. There are approximately ten known neurotransmitters that are classified in two different ways. The first classification is by molecular makeup where amino acids (e.g., glutamate, GABA), amines (e.g., epinephrine, dopamine, serotonin) and peptides (e.g., acetylcholine) are the dominate types. The second classification is by function, where some neurotransmitters induce depolarization (called *excitatory neurotransmitters*) and some induce hyperpolarizations (called *inhibitory neurotransmitters*). The action of one type of neurotransmitter may, however, be excitatory in one region of the nervous system while being inhibitory in other regions. It is important to note that there are many other molecules (e.g., vasopressin, neurotensin, insulin) and ions (e.g., zinc) that can impact synapses indirectly and are sometimes considered neurotransmitters. Furthermore, some neurotransmitters have other functional roles. For example, in a developing brain, neurotransmitters can act to guide and strengthen synaptic connections.

6.2   THE PRE-SYNAPSE

The release of neurotransmitter at the pre-synapse is commonly modeled in one of three ways that vary in complexity. The most complex is to model the binding of Ca^{2+}, activation of the vesicle, and release as separate steps. An example is the kinematic model

$$4Ca_i + X \underset{k_u}{\overset{k_b}{\rightleftharpoons}} X^*$$

$$X^* + V_e \underset{k_2}{\overset{k_1}{\rightleftharpoons}} V_e^* \overset{k_3}{\longrightarrow} nT$$

where X and X^* represent inactive and active Ca^{2+} binding proteins, V_e and V_e^* represent inactive and active vesicles, T is a molecule of neurotransmitter, and n is the number of neurotransmitters in a vesicle. Note that it takes four Ca^{2+} ions to activate a single binding protein, a number that has been observed experimentally. There are much more complex models of the Ca^{2+} modulated release of neurotransmitter that include ion channels for Calcium influx as well as pumps to return Ca^{2+} to the synaptic cleft. These models are often similar in form to I_K and I_{Na} discussed in Ch. 3.

A simpler model can be created by defining an equation for neurotransmitter molecules released as a function of the pre-synaptic voltage. A typical function is the Bolzmann distribution:

$$T(V_m^{\text{pre}}) = \frac{T_{\max}}{1 + e^{-(V_m^{\text{pre}} - V_p)/k_p}} \tag{6.1}$$

where $T_{\max}$ is the maximum concentration that can be released, V_m^{pre} is the pre synaptic voltage, V_p is called the half activation potential, and k_p is a slope factor. A third simplification will be discussed in Sec. 6.6.

6.3   NEUROTRANSMITTER DIFFUSION AND CLEARANCE

The diffusion of neurotransmitter across the synaptic cleft is typically not modeled. Instead, it is assumed that the released T molecules raise the concentration in the tiny cleft faster than the reaction time of the pre or post synapse. Neurotransmitter *clearance*, on the other hand, may be relatively slow. Clearance is accomplished either by reuptake of the molecule by the pre-synapse or by an enzyme that reverts the neurotransmitter to an inactive state. Both of these mechanisms can be modeled using differential equations, the most simple of which is a first order decaying exponential.

$$\frac{dT}{dt} = -k_t T \ . \tag{6.2}$$

Typically, the neurotransmitter (either in an active or inactive form) is taken back into the pre-synaptic axon and transported back to the soma. Here the molecules are reassembled, repackaged into vesicles, and sent back to the pre-synapse for future release. It is worth noting that if a neuron is undergoing bursting it is possible for release and uptake of a neurotransmitter to be occurring at the same time.

6.4   THE POST-SYNAPSE

6.4.1   The Post-Synaptic Current

The post-synapse translates a concentration of neurotransmitter in the cleft into a change in the post-synaptic potential (V_m^{post}). The neurotransmitter binds to a post-synaptic docking site, a special protein, and either directly or indirectly opens or closes ion channels embedded in the post-synaptic membrane. The current generated by the ionic flow across the post-synaptic membrane is defined the same way as other nonlinear membrane currents.

$$I_{\text{syn}}(t) = g_{\text{syn}} O(t) \left[V_m^{\text{post}} - E_{\text{syn}} \right] \tag{6.3}$$

where V_m^{post} is the post-synaptic membrane potential, g_{syn} is the maximum conductance, O is the probability of an open channel, and E_{syn} is the reversal potential. Although most post-synaptic currents are carried by one ion species, it is customary to refer to a current by the neurotransmitter which causes the channel to open. For example, a NMDA-gated post-synaptic current would be referred to as I_{NMDA}.

6.4.2   Excitatory, Inhibitory and Silent Synapses

Given the formulation of Eq. (6.3), E_{syn} will determine whether the synapse will be *excitatory*, *inhibitory*, or *silent*. To classify E_{syn}, we will assume that the post synaptic potential is at rest ($V_m^{\text{post}} = V_m^{\text{rest}}$). If $E_{\text{syn}} > V_m^{\text{post}}$, I_{syn} will be negative. Remember that a negative membrane current will depolarize the membrane so I_{syn} will *excite* the cell membrane. The resulting depolarization is called an *Excitatory Post Synaptic Potential* (EPSP). Likewise, if $E_{\text{syn}} < V_m^{\text{post}}$, I_{syn} will be positive, hyperpolarize V_m^{post} and *inhibit* the cell membrane. The hyperpolarization of the post-synapse is called an *Inhibitory Post Synaptic Potential* (IPSP). If $E_{\text{syn}} = V_m^{\text{rest}}$ the synapse is said to be *silent*. The role of the silent synapse will be explored further in Sec. 6.5.

6.4.3   Neurotransmitter Gating

The gating mechanism for a synaptic current, O in Eq. (6.3), may be described in the same way as the gating of any other ionic channel (see Ch. 3). The exception is that instead of being depen-dent upon the membrane voltage, synaptic gating is typically dependent upon the concentration of neurotransmitter in the synaptic cleft. The opening and closing of a channel which is dependent upon a concentration and not voltage is often called a *ligand-gated* channel. There are many possible formulations for O so we will cover only the most significant. A generic neurotransmitter gated reaction is

$$C + nT \xrightleftharpoons[\beta]{\alpha} O$$

governed by a differential equation

$$\frac{dO}{dt} = \alpha[T](1 - O) - \beta O \tag{6.4}$$

where $[T]$ is the concentration of neurotransmitter present. Similar to other gating mechanisms we could rewrite Eqs. (6.4)

$$\frac{dO}{dt} = \frac{O_\infty - O}{\tau_O} \tag{6.5}$$

$$O_\infty = \frac{\alpha T_{\max}}{\alpha T_{\max} + \beta} \tag{6.6}$$

$$\tau_O = \frac{1}{\alpha T_{\max} + \beta} . \tag{6.7}$$

where T_{max} is the concentration of neurotransmitter which will open all post-synaptic channels. More simply, $O(t)$ could be described by an analytic function, for example a bi-exponential

$$O(t - \tau_s) = K \left[e^{-(t-t_s)/\tau_2} - e^{-(t-t_s)/\tau_1} \right] \tag{6.8}$$

where K is the maximum amplitude of O (usually 1), τ_s is the time of the presynaptic spike, τ_1 is the rising time constant, and τ_2 is the falling time constant.

6.4.4   Multiple Gating Mechanism

It is possible for a synaptic current to be gated by a more complex combination of neurotransmitters, V_m^{post}, ionic concentrations and secondary chemicals. As an example, the model below is for the NMDA synaptic current that is dependent upon $[T]$, the extracellular concentration of magnesium ($[Mg^{2+}]_e$), and V_m^{post}:

$$I_{NMDA} = g_{NMDA} \cdot B \left[V_m^{\text{post}}, [Mg^{2+}]_e \right] \cdot O(t)[V_m^{\text{post}} - E_{NMDA}] \tag{6.9}$$

where $O(t)$ has a similar formulation, and dependence on $[T]$, as Eq. (6.4) and

$$B \left[V_m^{\text{post}}, [Mg^{2+}]_o \right] = \frac{1}{1 + exp \left[\frac{-(0.062 V_m^{\text{post}})[Mg^{2+}]_e}{3.57} \right]} . \tag{6.10}$$

6.4.5   Second Messenger Gating

Although we will not show the differential equations, the following reaction illustrates how a G-protein second messenger may be used in the gating of the $GABA_b$ receptor

$$R_o + T \rightleftharpoons R \rightleftharpoons D$$

$$R + G_0 \rightleftharpoons RG \rightleftharpoons R + G$$

$$G \rightleftharpoons G_o$$

$$R \rightleftharpoons R_o$$

$$K^+ + nG \rightleftharpoons O$$

First, $[T]$ binds to a free receptor R_o, which becomes active (R). At some later time, R can become deactivated (D). An inactive g-protein (G_o) can bind with R to become active (G) through a two step process. n activated g-proteins can then bind with a single K^+ ion to open the channel. In the background, the active R and G are inactivated (reset) and ready for another reaction.

6.5   SYNAPTIC SUMMATION

Above, we modeled I_{syn} as a single generic synaptic current. In reality, a single post synapse may have more than one type of receptor, even mixing excitatory and inhibitory receptors. To model a synapse with multiple types of receptors we can sum up the effects as in the parallel conductance model.

$$I_{syn}(t) = I_{syn}^{GABA} + I_{syn}^{Dopamine} + I_{syn}^{Serotonin} + I_{syn}^{silent} \tag{6.11}$$

where I_{syn}^{GABA}, $I_{syn}^{Dopamine}$, $I_{syn}^{Serotonin}$, and I_{syn}^{silent} are formulated using some form of Eq. (6.3). In this way, one synapse can respond to many different molecules.

Given this summation we can now understand the role of the silent synaptic current. As a silent synapse has a reversal potential at the resting potential, it will try to depolarize the membrane if $V_m^{post} < V_m$ and hyperpolarize the membrane if $V_m^{post} < V_m$. In other words, the direction of I_{syn}^{silent} will be such as to bring the post synaptic potential back to rest. This is a very important feedback mechanism since it prevents the potential from deviating far from rest. You can also think of the silent synapse as a type of inertia that the excitatory and inhibitory receptors must overcome to be effective.

Three additional properties of synapses are important. First, it is also possible for a post-synapse to form directly on the soma in which case there will be no attenuation due to the dendrites.

Second, the left-hand side of the reactions in Sec. 6.4.5 reveal how dependent an open channel is on all of the reactions. For example, if there are not enough inactive g-proteins (G_o), then no more channels can open. This is a problem called *availability* and is a rate-limiting step in any step-wise opening of the post-synaptic channels. Third, the opposite problem may occur if $O \rightarrow 1$, meaning that all the post-synaptic channels are open. When this is the case the post-synapse has become *saturated* and additional neurotransmitter will not have any effect.

6.6   SIMPLIFIED MODELS OF THE SYNAPSE

While kinematic and function-based models attempt to capture some aspects of the physiology, two related and more simple models are sometimes used in computer simulations. First, if an action potential is detected at the pre-synapse, a short pulse of T will be released into the synaptic cleft. Second, the steps 5–10 of synaptic transmission may be replaced by a simple time delay. In other words, if an action potential is present at the pre-synapse it will directly cause a depolarization or hyperpolarization, after some delay, in the post-synapse. The simplification involves removing the neurotransmitter and creating a direct electrical connection between the pre and post synapse.

Other simplifications will replace the differential equation based formulations with a simpler model. For example, the conductance in Eq. (6.3) may be modeled as an *alpha function* (top panel of Fig. 6.2).

$$G(t) = g_{max} \frac{t}{\tau} e^{1-t/\tau} \tag{6.12}$$

$$I_{syn} = G(t) \left[V_m^{post} - E_{syn} \right] \tag{6.13}$$

where t is the time after the action potential reaches the axon terminal, τ is a time constant, and g_{max} is the maximum possible conductance. A slightly more complex model is a bi-exponential (bottom panel of Fig. 6.2)

$$G(t) = \frac{A g_{max}}{\tau_1 - \tau_2} \left(e^{-t/\tau_1} - e^{-t/\tau_2} \right) . \tag{6.14}$$

An even more simple model would be to skip the conductance term all together and simply model the synaptic current, I_{syn}, as a pulse that follows the arrival of a pre-synaptic action potential. In any of these simple synapse models it is possible to control the onset time and therefore easy to introduce any delays that may occur.

6.7   THE MANY MECHANISMS OF DISEASES AND DRUGS

The effect of many drugs and diseases in the brain can be generalized as either decreasing or increasing the effect that the pre-synaptic potential has on the post-synaptic potential. There are, however, many methods by which these effects may be achieved. Below are five different mechanisms.

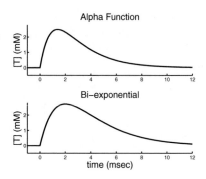

Figure 6.2: Alpha function and bi-exponential model for synaptic conductance.

1. If the neurotransmitter uptake mechanism is not present due to a mutation or is blocked by a drug, the neurotransmitter will not be cleared and continue to have an impact. Alternatively, the uptake could be too effective and the neurotransmitter will be cleared before it can have its full effect.

2. If drugs or mutations either block Ca^{2+} channels or bind to Ca^{2+} ions (a buffer) in the pre-synapse, the mechanism for vesicle release will be compromised. The impact is that no neurotransmitter will be released into the cleft.

3. The structure of the neurotransmitter itself could be compromised either by a mutation or by a drug that binds to the neurotransmitter before it can reach the post-synapse.

4. The post-synaptic channel could be compromised either by disallowing neurotransmitter binding (e.g., docking proteins on the I_{syn} channel are mutated) or by a change in the channel itself that limits the passage of ions. A more complicated target for drugs or diseases is to alter a second messenger which relays the message of neurotransmitter binding.

5. Diseases and drugs can induce long-term changes on the number of vesicles available, number of neurotransmitters per vesicle, or number of post synaptic binding sites. These changes may impact availability and saturation.

Most drugs have more than one effect because the same drug can impact different regions of the brain in different ways. Furthermore, the most powerful drugs often act through more than one of the mechanisms on the list above.

6.8   SYNAPTIC PLASTICITY AND MEMORY

In 1949, Donald Hebb published a book, *The Organization of Behavior*, which introduced the idea that the "strength" of a synapse could change over time in response to external factors. His idea was that *local memory* could be encoded in the strength of the connections between neurons. By strength, Hebb meant that the pre-synapse could have a greater impact (either excitatory or inhibitory) on the post-synaptic membrane. Given all of the ways that drugs and diseases can act on the synapse, it is possible for natural changes in synaptic strength to be accomplished by any or all of these mechanisms.

The ability of a synaptic strength to change is called synaptic *plasticity*. The mechanism of plasticity is one of the most elusive questions in neurobiology because of the many factors that could play a role. What is known is that some molecular level change occurs in either the pre-synapse, post-synapse or neurotransmitter clearance. For example, if a receptor channel on the post-synapse is over expressed, i.e., more channels waiting to bind with a neurotransmitter, the post-synapse will become more sensitive to the concentration of neurotransmitter and be less susceptible to saturation. The most simple way to model synaptic plasticity is to change g_{syn} in Eq. (6.3). The interpretation is that the value of g_{syn} is related to the number of synaptic ion channels. The opposite is true if the receptor channel is under expressed.

Hebb also proposed a mechanism by which synapses would be triggered to change, elegantly expressed in the phrase *neurons that fire together are wired together*. In other words, the more two neurons fire at the same time, the more tightly they will be connected. These types of synapses were in fact found to exist and were given the name *Hebbian synapses*. Recent studies, however, have shown that there are other reasons that neurons may either connect or disconnect, including pH, ion concentration, or drug use.

One of the most interesting ways to elicit changes in synapses is the phenomenon of *Long Term Potentiation* (LTP) discovered by Jerje Lomo in 1966. A typical experiment has four phases that are shown in Fig. 6.3.

First, an external stimulus is applied to one neuron, which fires an action potential. This action potential releases neurotransmitter which crosses the synaptic cleft and induces a depolarization in the post-synapse of a second neuron. The recorded EPSP, V_1^{post} serves as a baseline. Second, a rapid train of stimuli are applied to the first neuron. The number of stimuli and rate is variable but is typically performed over a period of minutes. Third, at some later time a single stimulus is applied to neuron one and V_2^{post} is recorded. Surprisingly, the response has been *potentiated*, or $V_2^{post} > V_1^{post}$. The interpretation is that the post-synapse has become more sensitive to activity at the pre-synapse, or in other words, the connection has become strengthened. Even more surprising is that this effect may last many weeks or even months. Understanding the mechanism of LTP is a very active area of research since is a controlled way to study how synaptic strength changes.

Summary

(1) Synapses and gap junctions form connections between neurons.

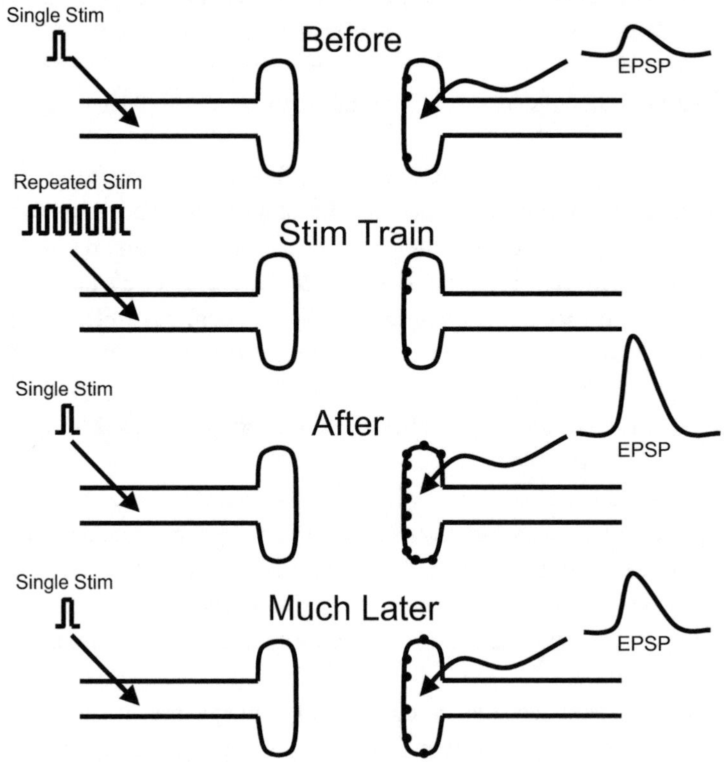

Figure 6.3: Long-term potentiation of EPSP.

(2) The pre-synpase converts an electrical impulse into a chemical signal. Models of the pre-synapse relate transmembrane voltage and Calcium concentrations to the release of Neurotransmitter into the synaptic cleft.

(3) The post-synpase converts a chemical signal into an electrical signal. Models of the post-synapse are ion channels that depend upon the concentration of neurotransmitter in the synaptic cleft.

(4) Enzymes act within the cleft to clear neurotransmitter.

(5) There are many steps involved in the transmission of information across the synapse and not all steps are known.

(6) Many diseases are a manifestation of a slight change in this transmission.

(7) Many drugs act by mimicking the dynamics of naturally occurring neurotransmitters.

(8) Local memory and learning occur at the synapse as changes in the number of receptors on the post-synapse.

CHAPTER 7

Networks of Neurons

It is often assumed that an individual neuron does very little in the way of high level information processing or learning. One hypothesis is therefore that it is the complex connections and interactions between neurons that give rise to the enormous range of functions and behaviors of the brain. The first half of this chapter will focus largely on the work of Roger Traub's research group on simulating thousands of multicompartment neurons. The second half of this chapter will introduce some basic concepts of artificial neural networks.

7.1 NETWORKS OF NEURONS

Previous chapters built up the theory necessary to create a multicompartment neuron with any complex geometry as well as the connections between neurons at the synapse. This section will demonstrate how many neurons may be connected together to perform some basic functions of networks of neurons.

7.1.1 The Traub Pyramidal Neuron

Based upon experimental studies in slices of the rat hippocampus, Traub and Miles created a generic Pyramidal neuron. A schematic of the neuron is shown in Fig. 7.1 and is composed of 28 compartments, divided into apical and basilar dendrites separated by the soma.

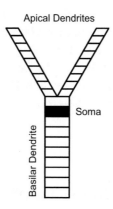

Figure 7.1: Traub pyramidal cell geometry.

The dendrites are modeled as passive cables and the soma is modeled using the bursting membrane model described in Sec. 3.4.2. For a complete description of the model, see *Neuronal Networks of the Hippocampus* by Traub and Miles (1991).

7.1.2   Conduction down the Axon and Synaptic Connections

The axon of the Pyramidal cell is not modeled directly because it is assumed that any action potential at the soma will reach the axon terminal after some delay. The axon is therefore modeled simply as a delay that depends upon the conduction velocity and length of the axon.

Once an action potential is delayed and reaches the pre-synapse, the impact on the post-synapse is modeled using the generic synaptic current equation first introduced in Eq. (6.3).

$$I_{\text{syn}}(t) = c_e O(t) \left[V_m^{\text{post}} - E_e \right] \tag{7.1}$$

where the maximum conductance term is c_e and the gating variable, O, takes the form

$$\frac{dO}{dt} = \text{Rect}(t_{\text{on}}, t_{\text{off}}) - O \ . \tag{7.2}$$

The Rect function is a pulse that is equal to 1 between t_{on} and t_{off} and 0 everywhere else. In Traub's model, if the soma voltage is greater than $20 mV$ and hasn't fired in $3 msec$ (i.e., refractory period), then t_{on} is set to the delay for the axon. t_{off} determines how long the pulse will last and is governed by the type of synapse.

7.1.3   Interneurons

Real networks of neurons are rarely composed of a single cell type. Traub's group therefore incorporated *interneurons*, also known as inhibitory neurons, into their model by making three changes to the Pyramidal cell model. First, the multicompartment geometry was reduced. Second, the active membrane at the soma was changed to eliminate the Calcium current. Third, synaptic connection from interneurons were made to be inhibitory by decreasing the reversal potential below rest and increasing t_{off} in Eqs. (7.1) and (7.2) at the post-synapse. The strength of the inhibitory connection was called c_i.

7.2   MODEL BEHAVIOR

Traub's research lab was able to simulate nearly 10,000 individual cells and made some ground-breaking observations that agreed well with experimental recordings. Before explaining the large-scale trends, they demonstrated three interesting phenomenon, in networks of two or three cells, that formed the basis for the more complex behavior. We repeat these simulated experiments below.

7.2.1   All or None Synaptic Induced Firing

The strength of an excitatory synapse may be modulated through c_e and was found to initiate an action potential burst in an all-or-none fashion. Figure 7.2 shows Pyramidal cell 1 connected to Pyramidal cell 2 through an excitatory synapse at the soma of cell 2. A burst was initiated at the soma of cell 1 and the strength of the synapse connection (c_e) was varied. It is clear from the figure that bursting in cell 2 is a governed by the strength of the connection.

In large scale models, it is difficult to always draw the shape of every neuron, so many presentations of *neural circuits* use some combination of circles to represent cells and triangles to represent synapses. In Fig. 7.2, the excitatory connection between two Pyramidal cells is represented by two open circles (e.g., excitatory cells) connected together by a line terminating in an open triangle (excitatory synapse). The direction of the triangle also indicates that cell 1 is driving cell 2.

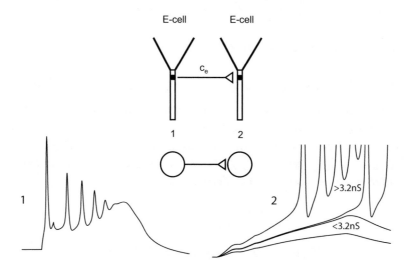

Figure 7.2: All-or-none synaptic transmission of bursting.

7.2.2   Suppression of Neuronal Firing

The bursting in the second neuron may be inhibited by an interneuron. Figure 7.3 shows the case where Pyramidal cell 1 will tend to cause Pyramidal cell 2 to fire (e.g., $c_e > 3.2nS$) in the absence of inhibition. Interneuron cell 3 (closed circle), on the other hand, will tend to prevent Pyramidal cell 2 from firing. Here it is the strength of the inhibitory connection (c_i, closed triangle) that determines whether cell 2 will fire or not. For example, if $c_i > 5nS$ the strength of the inhibitory connection will prevent an action potential from firing in cell 2. If $c_i < 5nS$, the inhibitory connection is not strong enough to prevent an action potential in cell 2, but it may delay when cell 2 fires.

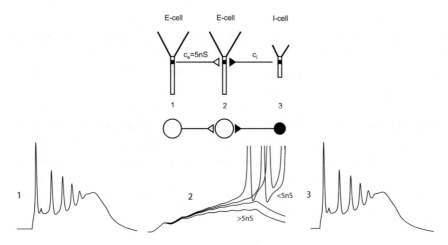

Figure 7.3: All-or-none inhibition of bursting by strength of inhibition.

7.2.3   The Importance of Delays

The impact of an inhibitory cell is governed not only by the strength of the connection, but also the timing. Figure 7.4 shows a similar three cell network as Fig. 7.3, but c_e and c_i have been fixed so that cell 2 does not fire. The timing of the interneuron action potential, however, was delayed relative to the firing of Pyramidal cell 1. As this delay becomes large, Pyramidal cell 1 has already committed to bursting and the late inhibition has little effect.

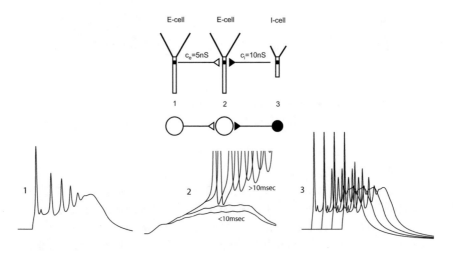

Figure 7.4: All-or-none inhibition of bursting by delays.

7.2.4   Central Pattern Generators

The brain is known to generate a number of rhythms that vary depending upon the waking state and mental tasks being performed. Computational and theoretical work have demonstrated that the brain may use the same neural circuits to generate different rhythms. Typically, these circuits consist of tens of thousands, or even millions of cells. As a simple demonstration, consider the left-hand panel of Fig. 7.5. The single pulse may activate Cell 1, which will activate Cell 2 and on to Cell 3. If the delay for passing the electrical impulse through the network is long enough, Cell 3 could then reinitiate firing in Cell 1 after it has repolarized. In this way, a simple neural circuit, possibly composed of many cells, could have some natural period of firing.

To modulate the rate of the cycle, consider adding an interneuron, as in the right panel of Fig. 7.5. If c_i is high, it may end the cycle all together. However, c_i could also be a way to delay how quickly Cell 3 reinitiates firing in Cell 1, thus slowing down the rate of the cyclical firing. This situation could be even more interesting if c_i itself was modulated by other synapses, hormones or ionic concentrations.

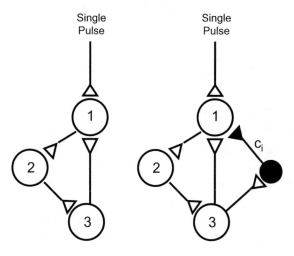

Figure 7.5: Cyclic firing in a network of neurons.

Groups of neurons that generate cyclic activity are often called *central pattern generators* (CPGs). As in Fig. 7.5, a group of neurons has an external stimulus that acts as a *command signal*, which can be used to turn on or off the oscillation. In some cases the loop may take in several command signals, some which turn the cyclic activity on and off and others which control the rate at which electrical waves propagate around the neural circuit. Most often the CPG will operate within some range of frequencies.

One very important application for CPGs is the coordination of muscles. The various outputs of a CPG can be used to create a pattern of signals that are sent to muscles. As a simple example, consider sending the output of Cell 1 to the left calf, the output of Cell 2 to the left quadriceps and the output of Cell 3 to the left hamstring. By tuning the delays in the propagation of waves around this circuit, we can control the coordination of muscle to take a step. Furthermore, we can imagine a similar circuit which controls the right leg. These two loops (right and left) can be connected together with inhibitory connections such that when the left Cell 1 is firing, the right Cell 1 is prevented from firing. Similar inhibitory connections can be made between the other cells in the left and right loop. Again, if the timing is tuned, the two legs can be at different cycles of a step and give rise to walking. Similar processes are involved in other coordinated cyclic activities such as respiration.

A second important application for CPGs is the coordination of mental activities. If we suppose that some task, such as adding two numbers together, requires a series of neuronal circuits to be fired in a precise sequence, a CPG can coordinate this activity. Similar cyclic processes may apply to the storage of short term memories in the hippocampus and the coordinated transfer of short-term memories to long-term storage in the cortex. Furthermore, the recall of long-term memories may involve the coordinated retrieval of information from different populations of neurons in the cortex.

7.3 ARTIFICIAL NEURAL NETWORKS

In 1943, McCulloch and Pitts published an article that showed how the basic actions of a neuron could be generalized. Furthermore, they showed that these general units, called *perceptrons*, could be used to perform logical functions such as *AND* and *OR*. Each perceptron is a system with a simple input/output transfer function that roughly models the behavior of a neuron. Because of the simple transfer function, many neurons (thousands or even millions) may be simulated quickly. The general philosophy, unlike much of this text, is that the higher-level functions of the brain are governed by the *connections* between neurons and not the neurons themselves. So, to model the higher-level functions of the brain (learning, distributed memory, planning, pattern recognition), it is not necessary to model the details of the neuron. Rather, it is more computationally efficient to model neurons as simple units that are connected together in complex ways as in Fig. 7.6. The claim is that it is large groups of simple neurons that perform interesting high level behavior such as learning, pattern recognition and planning. In the following section we will give a brief overview of neural networks composed of perceptrons.

7.3.1 The Perceptron and Single Layer Networks

In general, a perception, j, may only be in one of two *states*, firing (1) or not firing (0). This state is the output of the perceptron, a_j. But the perceptron may also take as inputs (a_i), the output of other perceptrons. Computing the output of a perceptron given the inputs is a two step process. First a weighted sum of the inputs is computed as

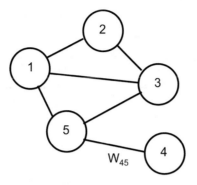

Figure 7.6: Weighted connections between perceptrons.

$$\sum = \sum_{i=1}^{N} w(i, j)a_i \tag{7.3}$$

where a_i are the N inputs and $w(i, j)$ are weights between perceptions i and j. The second step is to apply an *activation function*

$$a_j = g\left(\sum\right) = g\left(\sum_{i=1}^{N} w(i, j)a_i\right). \tag{7.4}$$

where a_j is the output. The function $g()$ may be a hard threshold as in the left panel of Fig. 7.7 and defined by

$$a_j = g(\sum) = \begin{cases} 1 & \text{if } \sum \geq \theta \\ 0 & \sum \leq \theta \end{cases}$$

or a more smooth function such as a sigmoid

Using the threshold $g()$ function and appropriate choice of weights, McCulloch and Pitts were able to show how to create **NOT, AND,** and **OR** logical functions as shown in the top panel of Fig. 7.8. The input to the gates (e.g., a_1 and a_2) are always either 0 or 1. Note that the only change needed to create different gates are the thresholds and weights. For example setting both weights to be 1 and the threshold to be -1.5 results in an **AND** gate.

As shown in the bottom panel of Fig. 7.8, the logic function can also be considered graphically. Inputs are graphed on the axes and the squares represent the desired outputs, filled for a logical 1 and open for a logical 0. The dotted line represents a *decision line* that is determined by the weights and threshold.

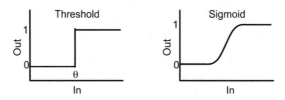

Figure 7.7: Threshold and sigmoid activation functions.

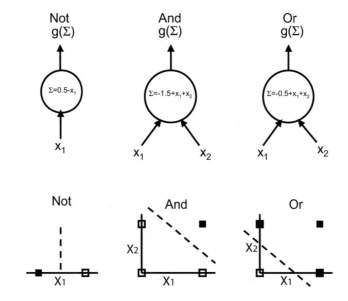

Figure 7.8: Perceptrons computing logical NOT, AND, and OR.

For a two-input perceptron, consider a group of inputs and weights that when summed are exactly at the threshold

$$\sum_{i=1}^{2} w_i a_i = \theta = w_1 a_1 + w_2 a_2 \ . \tag{7.5}$$

If we solve for a_2

$$a_2 = -\frac{w_1}{w_2} a_1 + \frac{\theta}{w_2} \ . \tag{7.6}$$

Equation (7.6) is in the form of a decision line, with a slope of $-w_1/w_2$ and intercept of θ/w_2, that separates the inputs into two categories. These two categories are assigned the output of 1 or 0 depending upon whether they are above or below the decision line. A decision line is therefore the most simple form of pattern recognition. Figure 7.8 shows the decision lines for the AND, OR, and NOT gates.

Although the math will not be presented here, given some set of inputs that need to be classified, the appropriate decision line may be created as in the left panel of Fig. 7.9. A decision line is found through a simple *learning rule* by which the weights and thresholds are tuned. Graphically, this is equivalent to changing the slope and intercept of the line in an iterative process until the line effectively separates to two inputs. Furthermore, this simple method will work for gates with many inputs where the dimension of Fig.7.8 would be increased. In this increased dimension, the learning rule would be adjusting a hyperplane to separate the classes of inputs.

The right panel of Fig. 7.9b demonstrates a case where a simple line is not capable of separating two sets of points. Here, open squares represent inputs which should be classified together and closed squares are in another category. The situation becomes even more complex if there are more than two classification categories. The problem is that the points in the right panel of Fig. 7.9 are not *linearly separable*, meaning they can not be separated by a line. The most simple example of a logical function that is not linearly separable is the **XOR** function shown graphically in Fig. 7.10. The problem of building and trained neural networks capable of classifying nonseparable points nearly ended active research in neural networks in the 1970s.

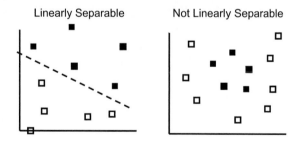

Figure 7.9: Example of linearly separable and not linearly separable data.

7.4 MULTI-LAYERED NETWORKS

The points in Figs. 7.9 and 7.10 can be classified if many lines are used. For example, to classify the XOR points, only two lines are needed. The practical meaning of adding more lines is the addition of *hidden layers* of perceptions in the network as shown in Fig. 7.11. Each unit receives inputs only from the preceding layer and can only send outputs to units in the layer ahead. Hidden layers allow

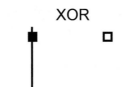

Figure 7.10: XOR logical function is not linearly separable.

for multiple types of points, separated by nonlinear regions, to be grouped together. Furthermore, adding more hidden layers, in principle allows for any complex set of regions to be defined.

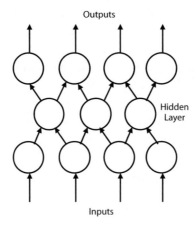

Figure 7.11: Generic feedforward multilayer network.

7.4.1   Backpropagation

Initially there was no way to systematically *tune* the weights to separate any generic set of groupings. Graphically, this is equivalent to changing the slope and intercept of many decision lines in a coordinated way. It was not until the early 1980's that a method was found. First, an ideal target output, T (a vector of outputs), is defined that corresponds to known inputs. Given an actual output, a_j, an error can be computed between the actual and target output. The outputs can then be adjusted by tuning the weights in a direction such that the sum of squared output errors, E_p, will be minimized.

$$E_p = \frac{1}{2} \left(T - a_j \right)^2 . \tag{7.7}$$

Substitution of Eq. (7.4) into Eq. (7.7) yields

$$E_p = \frac{1}{2}\left(T - \sum_{i=1}^{N} w(i, j)a_i\right)^2 . \tag{7.8}$$

Another interpretation of E_p is that it defines an *energy function* or energy surface. The surface is at a minimum when the error between a_i and T is smallest. In theory we might imagine the error being zero, however, in general it is not possible for the error to decrease below some finite value. Finding a practical solution is therefore equivalent to finding the set of weights that minimize Eq. (7.8). For an arbitrary case there is no analytical way to find the minimum. Instead the most common approach is the iterative method of *gradient descent*. First a function is defined that characterizes how the energy surface (E_p) will change if the weights (w) change.

$$\frac{dE_p}{dw} = \left[T - g\left(\sum w(i, j)a_i\right)\right] g'\left(\sum w(i, j)a_i\right) a_i \tag{7.9}$$

which is simply the derivative of Eq. (7.8) with respect to w. Evaluating Eq. (7.9) at a point gives the slope and direction of the energy surface with respect to w. The negative of this multidimensional slope is the direction w should be changed to cause the most rapid drop in E_p. This process is repeated in an iterative fashion to arrive at a minimum. It should be noted that for the gradient descent method to work, g must be a differentiable (e.g., a sigmoid). The new weights can then be computed as:

$$w^{\text{new}}(i, j) = w^{\text{old}}(i, j) + \alpha \frac{dE_p}{dw} \tag{7.10}$$

where α is called the *learning rate*. The updating of weights in this way is called the *delta rule* and is a specific case of a more general method called *backpropagation*. Backpropagation is a family of learning rules where blame is assigned to individual connection weights for errors. Once the blame is targeted, the weight may be moved in a direction that will decrease the error.

7.4.2 Momentum and Thermal Noise
The delta rule in Eq. (7.10) can be slow if α is small. On the other hand, if α is too large, the numerical solution may oscillate or even diverge from the minimum. A second *momentum* term may be added to Eq. (7.10):

$$w^{\text{new}}(i, j) = w^{\text{old}}(i, j) + \alpha \frac{dE_p}{dw} + v d w^{\text{old}}(ij) \tag{7.11}$$

where v is a momentum constant and dw^{old} is the previous change in w. Here, α may remain small, but the total change to w may be adjusted each iteration.

It is also possible for this iterative learning rule to converge on a local minimum and become stuck. To help reach a global minimum some neural networks use *simulated annealing*. Noise is added to Eq. (7.10), large at first but then slowly decreasing toward zero. The effect is that at some intermediate noise level, the current solution will not settle into any of the local minimums but the global minimum. A further decrease in the noise level, will not be enough to move the solution outside of the basin that forms the global minimum.

7.4.3   Training and Use

Most neural networks undergo two phases of use. The first we have already discussed and is called the *training* phase. It is during training that the weights of the network are adjusted by applying known inputs and specifying target outputs. After the network is performing well on the training set, the *usage* phase may begin. In other words, the network is ready for real and unknown inputs.

In more concrete terms, we may wish to create a neural network to recognize handwritten letters. To train the network we may present the letter 'E' written by several different people and adjust the weights until the network correctly identifies the *training set* of E's. The network might then be trained in the same way to identify other letters. In the usage phase we would present the network with letters from writers not represented in the training set. A good test would be to determine if the trained network can distinguish between an 'E' and an 'F'.

7.4.4   Energy Surfaces and Sheets

A multidimensional energy surface is not easy to visualize, but a three-dimensional analogy may make some concepts more clear. Imagine a sheet that is stretched across a number of chairs and tables. The sheet will naturally have some peaks that represent unstable points and valleys that represent local minima. If a ball is then dropped on the sheet, it will roll toward the nearest local minima and come to rest at the bottom. Here the topography of the sheet acts as an energy surface and the ball rolling toward the local minimum is similar to the method of gradient descent.

For the analogy to be useful, however, the minima should correspond to some desired output and the location where the ball is dropped should correspond to an input. Consider placing into each minima a perfectly scripted capital letter (e.g., 'E' in one minima, an 'F' in another minima). The initial location onto which the ball drops would then represent an imperfect letter that may be close to an 'E' but does not exactly match the perfectly scripted 'E'. In this case, the gradient descent method would classify the input as an 'E' by sending the ball into the local 'E' minima. Likewise, if the imperfect letter was closer to the ideal 'F', the ball would fall into the 'F' minima.

The analogy we have built up so far is for a network that has already been trained. To train our sheet-ball system, we would begin with a perfectly straight sheet and then present a perfect 'E' as an input. To ensure that the sheet would remember the 'E' in the future we would change the contour to make a local minima at exactly this point. The change in sheet topology would be analogous to changing the weights in a neural network. The effect would be to warp the sheet around the local minimum such that even a distorted 'E' input would roll to the lowest point. The same type of process could be used to generate other local minima to correspond to 'F', 'G', and 'H'. In the end, the sheet would be trained to recognize distorted letters.

In distorting the sheet to make room for the desired patterns, we may unintentionally create some local minima that do not correspond to any real letter. We would not want our ball to fall into one of these local depressions. To overcome this problem we can use the idea of simulated annealing. Physically, this would be the same as shaking the sheet as the ball rolls. The shaking would add a noise term and ensure that the ball rolled into one of the deeper minima that correspond to a real letter.

7.4.5   Network Structure and Connectivity

A consideration when building a neural network is how many perceptrons are needed to make the desired classification. If the network is too small, the error even at the global minimum may be unacceptable. The network in this case is an *underfit*. On the other hand, if the network is very large (an *overfit*), the increased computation may only result in a small gain in accuracy. A solution to this problem is to use *tiling*.

A tiling algorithm begins with a small network which is trained to classify inputs using the methods discussed above. When the global minimum is reached, the error is assessed. If the error is not acceptable, more perceptrons are added and new connections are made. The training then resumes on the larger network. This iterative process can be continued until the network produces results that are within a desired error tolerance.

7.4.6   Recurrent and Unsupervised Networks

Thus far, we have only considered *supervised* training of a neural network. Although supervision may have different interpretations, generally it means that the user specifies some known inputs paired with desired outputs, T. A learning algorithm, such as Eq. (7.10), can then be used to adjust the weights. In an *unsupervised* network, T may not be given and it is up to the network to develop classification patterns that pair similar inputs. The neural networks discussed thus far have been of a class called *feedforward* because information flows from one layer to the next in one direction. For a network to develop classification patterns, it must be able to adjust weights through feedback connections. Networks with built in feedback are often called *recurrent* networks. They can be trained to store associative memories and cycle through a sequence of patterned firings. The sequence of firings can be used to represent the coordination of muscle movements in a skill such as tying a shoe or throwing a ball. The two most mentioned types of recurrent networks are shown in Fig. 7.12 but the number of network variations has grown rapidly and is still growing. Although we will not discuss these types of networks, they have been shown to perform some incredible feats, such as unaided pattern recognition, storage of short-term memory and planning.

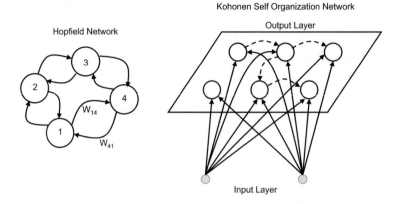

Figure 7.12: Hopfield and Kohonen recurrent networks.

7.5   NUMERICAL METHODS

Although it is possible to generate the computer code needed to simulate networks of neurons and artificial neural networks from scratch, a number of software packages exist which drastically reduce the start up time. To simulate networks of neurons, two programs have established themselves as the gold standards of neural compartment simulations: GENESIS and NEURON. GENESIS (http://www.genesis-sim.org) and NEURON (http://neuron.duke.edu) were born out of the need for a common platform for neural simulations. Both allow for stand-alone simulations on a single CPU or massively parallel simulations on supercomputers. Both have graphical user interfaces that show neural geometry, recording, and stimulating electrodes and built in functions for viewing the time courses of voltages, currents, and concentrations. To simulate artificial neural networks, MATLAB has a Neural Network tool box that contains many optimized functions for building, simulating visualizing, and analyzing neural networks.

Summary

(1) Networks of neurons may be studied using biophysical models, where the model of the neuron is relatively complex and performs a significant role in neural computation, or by artificial neural networks, where the model of neuron is relatively simple and it is the connections which perform a significant role in neural computation.

(2) In biophysical models excitatory and inhibitory synapses can be included, along with conduction delays. The advantage is increased realism. The disadvantage is increased computational cost.

(3) Simple neural circuits may be capable of exhibiting threshold behavior, a prerequisite for basic computation.

(4) Networks of neurons form central pattern generators which may direct the coordinated actions of other neural circuits or muscles.

(5) In artificial neural networks, the neuron and synapses have been greatly simplified. The advantages are that analytical analysis and defined learning rules are possible as well as relatively low computational costs and utility. The disadvantage is the potentially weak relationship between artificial neural networks and the actual processes occurring in the brain.

CHAPTER 8

Extracellular Recording and Stimulation

The theory we have derived so far is largely based upon transmembrane potentials and currents. Recording V_m or I_m in a real preparation, however, is nontrivial and nearly impossible in the clinic. Therefore, in most situations, it is an extracellular potential that is recorded. In this chapter we will use Maxwell's equations to derive a relationship between extracellular potentials and membrane currents.

8.1 MAXWELL'S EQUATIONS APPLIED TO NEURONS

Maxwell's equations describe how an electric field creates a distribution of electric potentials that can be recorded everywhere in space. One mechanism by which an electric field may arise is if a current source is present in a large conductive bath of fluid. In the physics and electrophysiology literature, a large bath is called a *volume conductor*. If we assume a small dipole current, I_o, emanates from a point in a uniform conducting medium of infinite extent, the current will flow radially in all directions. By placing a small sphere of radius, r, around the current source, we can compute the flux, J, through the sphere as

$$\mathbf{J} = \frac{I_o}{4\pi r^2}\mathbf{a_r} \tag{8.1}$$

where $4\pi r^2$ is the surface area of the sphere and $\mathbf{a_r}$ is a unit vector in the radial direction.

Although an electric field can theoretically change at the speed of light, biological systems are much slower. Therefore, all analysis of biological systems can assume changes in the electric field to be *quasi-static*. The practical meaning is that if we record a potential, even if it is far from the source, it will reflect the current at exactly that time. In reality, this is not true but the field changes much faster than the biological current can change. Given this assumption

$$\mathbf{E} = -\nabla\phi_e \tag{8.2}$$

where $\mathbf{E}$ is the electric field vector and $\nabla\phi_e$ is the gradient of extracellular potentials in the bath. To relate the gradient of ϕ_e to flux, we can use the general form of Ohm's Law:

$$\mathbf{J} = -\sigma_e\nabla\phi_e \tag{8.3}$$

where σ_e is the conductivity of the bath. Far from any current sources, Eq. (8.3) reduces to Laplace's Equation.

$$\nabla^2 \phi_e = 0 . \tag{8.4}$$

Combining Eqs. (8.1), (8.2), and (8.3),

$$\mathbf{J} = \frac{I_o}{4\pi r^2}\mathbf{a_r} = -\sigma_e \nabla \phi_e . \tag{8.5}$$

We know, however, that components of $\nabla \phi_e$ can only be in the radial direction, so

$$-\sigma_e \frac{d\phi_e}{dr} = \frac{I_o}{4\pi r^2} \tag{8.6}$$

integration with respect to r yields

$$\phi_e = \frac{I_o}{4\pi \sigma_e r} . \tag{8.7}$$

Note that in Eq. (8.7), ϕ_e is the same for any surface where r is constant (concentric spheres).

Equation (8.7) is useful when there is one current source. When more than one current source is present, we can sum up the effects of all the currents. First, we can assume each current is generated within some small volume, dV, and has some incremental effect on the overall ϕ_e. Differentiating Eq. (8.7)

$$d\phi_e = \frac{I_o}{4\pi \sigma_e r} dV \tag{8.8}$$

and to sum up the effects

$$\phi_e(x', y', z') = \frac{1}{4\pi \sigma_e} \int \frac{I_m(x, y, z)}{r} dV \tag{8.9}$$

$$r = \sqrt{(x - x')^2 + (y - y')^2 + (z - z')^2} \tag{8.10}$$

where r is the distance from each current source, (x, y, z), to the recording point, (x', y', z'). Note that we have assumed that the I_o of Eq. (8.8) is the membrane current, I_m, emanating from a small patch of membrane. In this way, we can theoretically calculate potentials in a bath surrounding an active neuron, given all of the I_m sources (see Fig. 8.1).

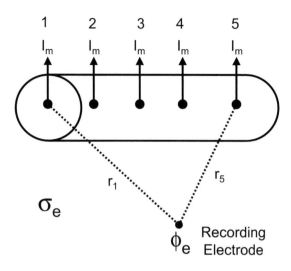

Figure 8.1: Summation of dipoles in calculating an extracellular potential.

8.1.1 Forward and Inverse Problems

If all of the I_m and r terms are known, ϕ_e may be calculated using Eq. (8.9). This calculation is known as the *forward problem* and has a unique solution. Equation (8.9) also demonstrates a fundamental problem encountered in all extracellular recordings. ϕ_e is an *average* of the effect of many current sources, weighted by the radii. As in any average, information is lost. Therefore, if you record ϕ_e there is no way to back calculate all of the I_m sources. In fact, you may not even know how many current sources compose ϕ_e. The computation of I_m sources from ϕ_e is called the *inverse problem*. Despite the limitations of the inverse problem, ϕ_e is much easier to record and is the typical measurement made for both research and clinical purposes. Although we will focus on the forward problem here, there exists a large body of research on gaining the maximum amount of information from ϕ_e. We will explore some of these methods in Ch. 9.

8.2 EXTRACELLULAR POTENTIAL RECORDINGS

8.2.1 Extracellular Potentials Generated by an Active Cable

In our derivation of propagation in a cable, we assumed that the cable was very thin. If we also assume that the cable is surrounded by a large bath and an electrode far from the cable, the current sources, I_m, will appear to be point sources. Furthermore, if we align the cable with the x-axis, then the infinitesimal volume in Eq. (8.1) becomes an infinitesimal length, dx:

$$\phi_e(x', y', z') = \frac{1}{4\pi\sigma_e} \int \frac{I_m(x, y, z)}{r} \, dx \tag{8.11}$$

where I_m at each point in the cable may be computed in one of two ways.

$$I_m = C_m \frac{dV_m}{dt} - I_{\text{ion}} \tag{8.12}$$

or

$$I_m = \frac{\pi r^2}{R_i} \frac{d^2 V_m}{dx^2} \, . \tag{8.13}$$

Note that in the second formulation, ϕ_e can be computed at a particular moment in time by knowing only the spatial variation in V_m along the cable.

8.2.2   Current Sources in a Complex Domain

Although a section of an axon or dendrite may be approximated by a straight cable, in reality axons and dendrites are not straight. Therefore, Eq. (8.11) is not valid. Furthermore, given the complex tangle of neurons, the assumption of an infinite bath of conductive fluid is not valid either. It is possible to numerically solve for ϕ_e in a complex domain by building a circuit analog as in Fig. 8.2. Although the circuit is only a very small portion of a much larger circuit network, it should be clear that we can represent intracellular and extracellular space by passive resistors and membranes as nonlinear current sources. The advantage of this arrangement, besides allowing for complex geometries, is that extracellular and intracellular potentials can be recorded directly at any point (e.g., Node 2). It also naturally allows for the membrane of the cell to influence currents and potentials in the extracellular space. At the boundaries we can make the simple assumption that no current may leave, leading to sealed ends as described in Sec. 4.4. To model a 3D section of neuronal tissue, it is conceptually simple to extend our resistor network into a third dimension.

The obvious disadvantage of a circuit network is that it is computationally very slow because potentials and currents must be computed at *every* point. The more subtle disadvantage is that the technology does not currently exist to efficiently image the complex web of neurons and directly translate the image to a circuit model.

8.3   THE ELECTROENCHEPHOGRAM (EEG)

From the previous section and Ch. 7 it should be clear that large populations of neurons, firing in some regular pattern, can generate a significant extracellular potential that may be detected at some distance. The more active and synchronized the neurons, the larger the signal. Of course, any behavior will be greatly averaged but a general idea of the degree of activity can be determined. At

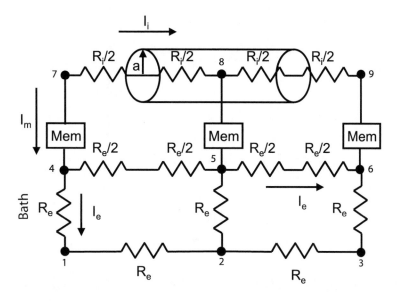

Figure 8.2: Extracellular potentials in a complex network of neurons.

least in principle, we have described precisely the situation in recording the *electroenchephalogram* or EEG. The EEG is the potential recorded from the surface of the scalp. The deflections, although in the μV range, correlate to the activity of the brain in regions under the electrode. Therefore, the EEG is typically recorded using many electrodes positioned directly above known brain structures.

8.4 PRACTICAL ASPECTS OF RECORDING NEURAL POTENTIALS

The typical process of recording and storing electric potentials is summarized in Fig. 8.3. In the section below, we review the key components of a recording system. Processing will be the subject of Ch. 9.

8.4.1 Electrodes

Currents in the brain are carried by ions. Currents in computers and wires, however, are carried by electrons. The role of an electrode is therefore to transduce ionic current to electric current. The chemical reaction most typically used is:

$$Ag \rightleftharpoons Ag^+ + e^-$$

$$Ag^+ + Cl^- \rightleftharpoons AgCl$$

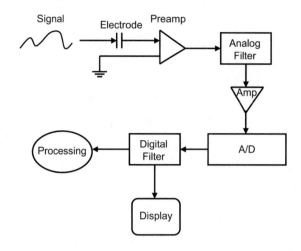

Figure 8.3: Recording, filtering, amplification, and digitization of an extracellular potential.

Although many metals could be used to create an electrode, the *Silver-Silver Chloride* electrode is easy to fabricate, has a fast electrical response, and is compatible with biological tissues.

Electrodes for recording potentials from tissue are typically small ($< 1mm$). They can be placed directly on the surface of neural tissue or, as neural tissue is somewhat spongy, can easily be pushed deep inside of a three-dimensional section of tissue. Recordings made in the brain are typically called *cortical recordings* if the electrodes are on the surface of the brain, or *sub-cortical* if they are pushed into the depth of the tissue. Sub-cortical recordings can in fact be made from awake patients because the brain has no sensory neurons and therefore causes no pain.

In EEG recordings we are not concerned with uncovering individual neuronal firings but rather the general activity below the electrode. Therefore, many electrodes are placed on the scalp so comparisons can be made between the relative activity of different brain regions. Since an average is desired, the electrodes often cover a large area (e.g., $1cm$ in diameter). The amplitude of the EEG is small because the skull and scalp have a low conductivity (σ_e). To decrease the impedance of ion flow to the electrode, the scalp may be *abraded* and a gel with free Cl^- ions is often applied. The most difficult problem in the interpretation of EEG is that it is an inverse problem. In other words, it is difficult to localize the current sources in the three dimensions of the brain from a few recordings on the two dimensional surface of the head.

8.4.2   Recording Preparations

Before discussing recording preparations it will be helpful to keep in mind the formal definition of a potential. The electric potential at a point is simply the work required to move a single charge (e.g.,

e^-, K^+) from infinity to that point against the electric field. Therefore, as $r \to \infty$, ϕ_e must equal zero.

In most tissue recordings there are at least three electrodes present. The *recording electrode* will detect the potentials resulting from the electric field generated by membrane currents. In the true definition of a potential, however, another electrode at infinity would be necessary. As this is impossible in practice, most recordings have a second electrode, called the *ground*, that is far from any electrical activity. Two common locations for the ground are a large slab of metal in the room or some location in the body that is very poorly electrically connected to the recording electrode, e.g. the skull. The third electrode is called the *reference* and is typically located in an electrically connected region of tissue but not close to the recording electrode. To understand the role of the reference, consider that the recording electrode will detect *all* current sources scaled by the radii. So, current sources close by will have a large impact but there will be few of them. On the other hand, the distant current sources will individually have a small contribution but there could be many of them. The role of the *reference* electrode is to record these distant sources so they may be subtracted from the recording signal.

8.4.3   Filtering, Amplification and Digitization

The amplitude from a neural recording (EEG or tissue recording) may be as small as $1 \mu V$ and contain noise from the recording device, room, and motion artifacts. Therefore, neural signals are generally filtered and amplified. In high-end recording systems, amplification and filtering are performed in a series of steps. Even after filtering and amplification, the signal is still continuous. To be stored in a computer, the signal must be *digitized*. Digitizing will transform a continuous signal (with infinitely many values) to a discrete signal (with some finite number of values). The specifics of how to amplify, filter, and digitize a neural signal will not be covered here but can be found in texts on biomedical instrumentation.

8.5   EXTRACELLULAR STIMULATION

If a stimulus is applied to one of the extracellular nodes in Fig. 8.2, the potential at that point (ϕ_e) will become more negative. According to Eq. (8.4), this negative potential will fall off with distance from the stimulus point. However, it is possible that even after the attenuation, the extracellular potential outside of an active section of membrane could be driven more negative. Recall that $V_m = \phi_i - \phi_e$ and that Sodium channels will open if $V_m > V_m^{th}$. So, if ϕ_e outside the membrane decreases enough, the membrane potential may reach threshold. In this way, an extracellular stimulus can be used to activate a patch of membrane.

In a real preparation, a stimulating electrode will be surrounded by a complex tangle of axons, dendrites, somas, and glial cells, so the exact amplitude and duration to induce an action potential will not be known. Furthermore, a single action potential propagating down a signal axon may not be enough to have any global impact, as it will be lost in the constant *chatter* of the neurons. Therefore, the goal of most neural stimulation is to cause all or most of the neurons around the stimulus electrode to fire and then recruit even more neurons. This idea of recruitment is called *capture*. To ensure that a given stimulus is achieving capture, it is common to record a potential at some distance from the stimulus, and then turn up the stimulus amplitude until the distant neurons are firing in response to the stimulus.

In addition to a stimulating electrode, a second electrode, called the *return*, is often used as well. The purpose of the return electrode is to provide a sink (or return) for the injected current. The advantage of using a return electrode is that current will find the shortest pathway(s) from the stimulus to the return. By placing the stimulus and return relatively close together it is possible to focus current into a particular location.

8.6   NUMERICAL METHODS: COMPUTING ϕ_E IN A CABLE

To compute $\phi_e(t)$ numerically, at each time step

$$\phi_e(t) = \frac{1}{4\pi \sigma_e} \sum \frac{I_m(t)}{r} \, dx \qquad (8.14)$$

must be computed.

```
Define Propagation and Membrane Constants
Define σ_e
Define location of recording electrode
Compute initial conditions
Compute radii (r(i)) for each patch to recording electrode.

for (time=0 to time=end in increments of dt)

    Imsum=0.0
    for (i=1 to i=Last Node in increments of (1)

        Update Currents
        Update Differential Equations
        Imsum = Imsum + (Im(i)/r(i))*dx

    end

    φ_e(time) =  1/(4π σ_e)*Imsum
    Save φ_e(time)

end
```

```
Store values of interest to a file
```

Summary

(1) Extracellular potentials may be modeled as a sum of transmembrane current dipole sources.

(2) The forward problem in electrophysiology is the computation of an extracellular potential given the dipoles. The forward problem may be solved exactly.

(3) The inverse problem in electrophysiology is the back calculation of the dipoles given an extracellular potential. The inverse problem is not mathematically well posed.

(4) Recordings from neurons require electrodes which will transduce ionic activity to current carried by electrons.

(5) Electrical signals recorded from real preparations are small in magnitude and noisy. Instrumentation is needed to filter amplify and digitize extracellular recordings for later storage and analysis.

(6) Extracellular stimulation can cause depolarization or hyperpolarization of distant cellular membranes.

Homework Problems

(1) Estimate $\phi_e(t)$ at the electrode in Fig. 8.4, where three current sources are propagating down a cable at $300mm/s$. Assume $dV = 2mm^3$, all distances are in mm, and all currents are in $\mu A/mm^2$.

(2) Derive the equations nodes 2, 5 and 8 in Fig. 8.2.

Simulation Problems

(1) Modify the program in Ch. 4 simulation problem 2 to compute $\phi_e(t)$ at $x = 1.5cm$ and $0.2cm$ above the cable. Assume that $\sigma_e = 1mS/cm$.

(2) Show how the shape of ϕ_e changes as the location of the recording electrode is changed. Explain the changes you see.

(3) Show how the shape of ϕ_e changes as R_i is changed. Explain the changes you see.

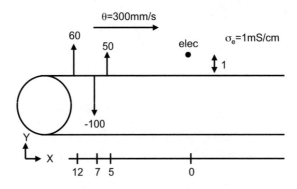

Figure 8.4: Conputing an extracellular potential.

(4) Program Fig. 8.2 assuming $R_e = 10\Omega cm$, $R_i = 100\Omega cm$, and I_{ion} is described by the Hodgkin-Huxely model.

(5) Use the program in problem 2 above to create a strength-duration curve for a stimulus at node 2 in Fig. 8.2.

CHAPTER 9

The Neural Code

Over the past several decades, two primary thrusts have driven neuroscience research. The first thrust has been to understand how neurons *encode* information and communicate with other neurons in the brain. Much of this text has been dedicated to this first thrust. An analogy would be to understand the physical devices that allow radio operators to use Morse code to communicate. The second thrust has been to *decode* neural messages by listening in on the chatter of neurons with extracellular electrodes. The analogy here is to learn Morse code and listen in a conversation between two radio operators. These two thrusts are by no means separate, and many researchers work in both areas. As the encoding and decoding of information in the brain is a rich field of study, we will only cover the most widely used techniques here. In this chapter, we will first examine how neurons use firing rates to encode information and then turn to the basics of how to decode the conversation.

9.1 NEURAL ENCODING

Experiments have shown that the shapes of action potentials do not vary significantly from neuron to neuron. It is assumed, therefore, that information is not transmitted from neuron to neuron encoded in the shape of the action potential. Neurons also do not fire at regular intervals and may even fire when no input is presented at their synapses. Therefore information is not transmitted in a single action potential. What does vary significantly is the natural frequency of firing when a neuron is excited versus when it is not excited. Furthermore, the firing rate, or frequency, of one neuron can be modulated by the synaptic inputs to that neuron. The firing of a neuron can also affect the firing rate of the neurons to which it is connected. A central principle of neuroscience is therefore that neural information is encoded in *firing frequency*.

9.1.1 The Membrane as a Frequency Detector

To understand how it is possible for neurons to encode information in the frequency of firings, consider a single current pulse applied to a membrane. In Sec. 2.3 we found that if a large stimulus was applied for a short time, the membrane may not fully charge to threshold. When the current was turned off, it required time for the membrane to discharge back to rest. Now, consider that a train of short-duration stimuli are applied to the membrane. If the duration between stimuli is long, the membrane will have time to fully discharge after each stimulus is turned off. If the stimulus interval time is decreased, the membrane will not fully discharge before the next stimulus is applied. The result is that the membrane potential will *drift* toward more depolarized potentials. This phenomenon is called *temporal integration* or *temporal summation*.

In Fig. 9.1, the total stimulus duration and amplitude remained constant and only the frequency was changed. The red plots show a low-frequency stimulus where the drift does not reach threshold. The black plots are for a higher frequency where the drift does reach threshold. These two different behaviors demonstrate how the frequency of stimulation is capable of modulating neural behavior at the cellular level. A secondary way that frequency can alter neural behavior is in the timing. Even when the rate is fast enough to induce an action potential to fire, the frequency can alter *when* the action potential fires. For example, in Fig. 9.1 a higher stimulus frequency caused an action potential to fire earlier.

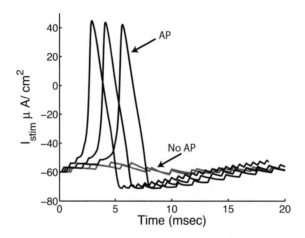

Figure 9.1: Effect of stimulus frequency on charging and action potential firing.

9.1.2   The Synapse as a Frequency Detector

A similar phenomenon to temporal integration of the cell membrane can occur at the synapse but at a slower rate. For example, if a number of action potentials reach the end of the axon terminal in rapid succession, short bursts of neurotransmitter will be released into the cleft faster than it can be cleared. Therefore, the concentration of neurotransmitter will slowly rise, leading to a sustained post-synaptic current. Recall that the post-synaptic current may either depolarize (excite) or hyperpolarize (inactive) the post synaptic membrane.

9.2   NEURAL DECODING

The purpose of neural decoding is to listen in on many neurons to understand how they are passing information to one another. As it is the timing of action potentials that is important, the extracellular potentials, ϕ_e, contain all the information needed to decode neural messages. From Ch. 8, we know that as an action potential approaches an extracellular electrode, ϕ_e becomes positive. As the action potential is moving away from the electrode, ϕ_e will become negative. Therefore, if an electrode is placed very close to an axon, the *biphasic* deflections in ϕ_e (as in Fig. 9.2) will reflect a single-action potential propagating down the axon.

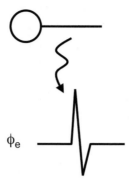

Figure 9.2: Example of a biphasic electrogram.

9.2.1 One Electrode, Many Recordings

When an extracellular electrode is very close to a single axon, other surrounding neurons contribute very little to ϕ_e and the recording will reflect only one neuron. In reality, neurons form a complex web and we would not know the exact location of our recording electrode within the web. Therefore, whenever one neuron fires and the action potential propagates down an axon, we would not know if the deflection we recorded is from one neuron or two neurons that fired at nearly the same time. Further complicating the extracellular potential time course is the fact that axons are not straight cables and may be at any distance or angle relative to the electrode. The result is that ϕ_e is no longer biphasic.

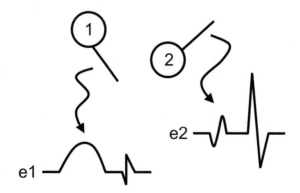

Figure 9.3: Electrograms from multineuron preparations.

A somewhat more challenging problem is to distinguish between the many neurons that may contribute to a single ϕ_e recording. Figure 9.3 shows the relatively simple case of two neurons being recorded by two electrodes. Notice that the deflections are not biphasic because of the orientation of the electrode relative to the axon. Furthermore, because of the geometry, the firing of cell 1 will appear as a different deflection at electrode 1 and electrode 2. This situation becomes even more complicated if many neurons are firing. Lastly, consider how difficult it would be to separate the firing of cell 1 and 2 if they fired at nearly the same time.

The key to neural decoding is to assume that the electrodes do not move and that the action potentials always propagate at the same speeds down their respective axons. With these assumptions the firing of cell 1 will always give the same pattern of deflections at electrodes 1 and 2. Likewise, the firing of cell 2 will always generate the deflections in ϕ_e. The deflection in ϕ_e corresponding to a particular neuron will therefore always have the same shape. We can then state that the first deflection in both recordings is due to cell 1 and the second deflection in both recordings is due to cells 2. Note, however, that we can not state that cell 1 *caused* cell 2 to fire. The situation would be greatly complicated if there were many participating neurons. In principle, however, it is possible to use the extracellular shapes to sort or *bin* the firings of individual neurons and therefore record the neural code from many neurons simultaneously.

9.2.2   Eavesdropping

An analogy to the recording and interpreting of neural information is to imagine entering a room filled with people speaking a foreign language that you do not understand. Upon entering the room you are far from every individual conversation and so you hear mostly noise. In fact, all you may be able to infer is how passionate the conversations are on the whole. If you move closer to any particular conversation you may be able to gain more information on that particular group. Again, you do not speak their language, but you may be able to gain some information about who is a leader, who is a follower, when arguments occur and when they are resolved. All the while, the noise is still in the background. Now, consider that you perform this task, day after day, week after week. Eventually, you would begin to recognize patterns in the language and pick up on subtleties of the conversations.

Eavesdropping on a foreign language class is similar to our current status in decoding neural patterns. We understand parts of the neural language, and can gain some idea of when a population of neurons is working on a problem, but on the whole they speak a language that is very foreign to us.

9.2.3   Spike Sorting

The prospect that a recording from one electrode contains information from many neurons is another step toward decoding neural information. The next step is to separate out the firing of individual neurons. Typically this separation is called *spike sorting*. The idea is to create a bin for each neuron and then sort the timing of firings into that bin. While there are many methods for accomplishing binning the most straightforward method is to define a number of *template deflections* that correspond to specific neurons firing (Fig. 9.4). This is an important step for two reasons. First, the number of template deflections will be the same as the number of bins and therefore is an assumption about how many neurons are contributing to ϕ_e. If there are too few or too many bins, some spikes will be misclassified. Second, the templates will serve as the ideal to which other deflections will be compared. Although the templates may be determined by eye, there do exist methods, such as Principle Component Analysis (PCA), which can automate the selection of templates using statistical differences in the signals.

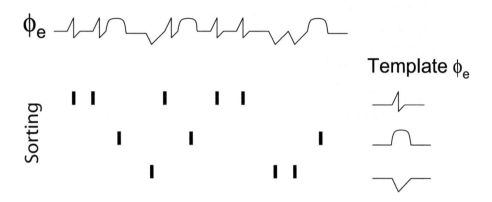

Figure 9.4: Spike sorting of extracellular deflections.

9.2.4   Time Binning and Firing Frequency
Once the timing of each neural firing has been extracted, the firing rate (frequency) can be found. It is often the case, however, that a neuron fires in a burst and then remains quiet, followed by another burst. To detect the change in frequency, the spikes may be binned in time. In *time binning*, all spikes are counted within a bin that span a given length of time. By counting the number of spikes that occur within the time bin, an average firing rate may be computed. Changes in firing frequency can therefore be tracked over time. For example, the sequence of firings in the right side of Fig. 9.5 have been separated into four bins.

9.2.5   A Graphical Representation
As extracellular potentials are relatively simple to record, it is often the case that many hundreds of electrodes are used. The result is hundreds of ϕ_e recordings from potentially thousands of neurons. After sorting and binning, information is often represented graphically in one of two ways. First, the individual spikes may be shown in a matrix, or *raster*, where each row represents a single neuron (Fig. 9.5, right panel). A raster plot shows all of the raw data but can be difficult to interpret. As information is encoded as a firing frequency, the timing of an individual spike is less important. The second graphical representation, therefore displays firing rate as a color (Fig. 9.5, left panel). Again, each row corresponds to a single neuron. The advantage of the color coded map is that the firing

frequency of different neurons may be correlated more easily. It is often the case that many trials may be recorded of the same phenomenon and the frequency results averaged before displaying the data graphically.

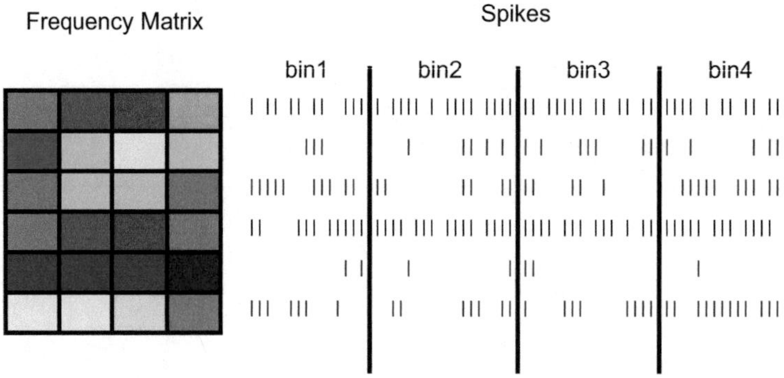

Figure 9.5: Graphical binning and frequency matrix.

9.2.6   Windowing

There are a number of potential problems with time binning. For example, consider the impact of slightly offsetting the four bins in Fig. 9.5. If the firing frequency changed suddenly, the change may be averaged out over two neighboring bins. A simple solution would be to add more time bins, each with a shorter duration. As the bin size is made smaller, however, there will be less possible spikes within each bin to count and the range of possible frequencies will be limited. On the other hand, if the bin size is large, the frequency curve will be much more discrete. A more clever method of binning is to wait for some number of spikes (e.g., 10 or 20) and count the time that elapses. In this way, the bin size adapts to the rate of firing. For fast firing the bin is small and for slow firing the bin expands. The problem with adjustable bin sizes is that it becomes difficult to correlate the firing patterns of different neurons.

A more dynamic method of computing firing frequency is to define a single bin of some duration (Δt) and slide it along the spike train. This technique is known as *windowing*. For each slide, the number of spikes can be counted and used to compute a frequency for that particular slide time. The simplest window may be defined as

$$w(t) = \begin{cases} 1/\Delta t & \text{if } -\Delta t/2 \leq t \leq \Delta t/2 \\ 0 & \text{Otherwise} \end{cases}.$$

If the slide is continuous rather than discrete then we can rewrite the frequency estimate as

$$f(t) = \int w(\tau)s(t - \tau)\, d\tau \tag{9.1}$$

where the window, w remains stationary and the spike train, $s(t)$, is moved. When presented in this way, the window is sometimes called a *kernel*.

It is also possible to use windows with different shapes that will weight some spikes more than others. For example, the Gaussian kernel is defined as

$$w(t) = \frac{1}{\sqrt{2\pi}\,\sigma_w} \exp\left(-\frac{\tau^2}{2\sigma_w^2}\right) \tag{9.2}$$

and so spikes closer to the evaluation time are more heavily weighted in the frequency estimate.

Thus far we have assumed that the entire spike train is known, and so times both before and after the evaluation time may be used. These kernels are called *noncausal* because they involve spikes that will occur after the evaluation time. In any real-time application, only spikes that occurred before the evaluation time are known and therefore a *causal* window must be used. In principle, the same window shapes can be used, but instead of evaluating at the center of the kernel the evaluation is performed at the right edge.

Given a number of electrodes, each recording from 10s of neurons, it is still not possible to definitely state which neurons are causing which other neurons to fire. Given a long enough recording, however, it is possible to build up strong correlations. For example, if we see that every time neuron 58 fires, neuron 238 fires with a $30msec$ delay, we can guess that they are either directly connected or connected through some other neurons. Likewise, we may observe that whenever neuron 58 is firing, neuron 189 is not, implying some direct or indirect inhibitory connection. In this way, a long recording from many neurons may be capable of mapping out the *functional connectivity* of a network of neurons.

9.3   TUNING CURVES

While Sec. 8.5 described how an extracellular stimulus may induce neural firing, the previous sections of this chapter described how to characterize a neural response. The combination of a controlled input (stimulation) and recorded output (neural firing rate) enables a transfer function to be created that characterizes a network of neurons. This type of transfer function is typically called a *tuning curve* and is represented graphically as a plot of a stimulus parameter versus firing rate. The stimulus parameter can be any number of possible input parameters. As there are too many possible types of stimuli we will broadly classify them as either direct electrical stimulation or indirect stimulation. Direct electrical stimulation can take the form of cortical or sub-cortical stimulation through electrodes or even magnetic stimulation (see Sec. 10.3.3). The stimulus parameters that may be varied are the stimulus strength, duration, or pacing rate. Indirect stimulation may take the form of light, smell, taste, touch, sound, hormones, or drugs injected into the blood stream.

Once the empirical data are collected, it is often the case that an analytic function, called a *tuning function*, is fit to the data. There are many mathematical forms for tuning functions but two deserve special mention here. First are a group of Gaussian-like curve that have a maximum firing rate at some particular value of the stimulus parameter with a lesser response for deviations from that value. Networks of neurons that have a Gaussian tuning curve have a sensitivity to a particular range of the stimulus. Second are a group of sigmoid-like functions that do not respond for some large range of stimuli, then rapidly transition to a strong response as the stimulus is changed. In other words, networks of neurons that have a sigmoidal tuning curve exhibit threshold behavior.

Figure 9.6 shows how cone cells, eye cells that detect color, respond to different wavelengths of light. The solid line shows the firing frequency of cones that detect red light. The dashed line shows cone cells that detect green light. The dotted line shows cone cells that detect blue light. Note that each type of cell has a background firing frequency but they will respond most strongly, i.e., increased firing rate, in response to a small range of wavelengths.

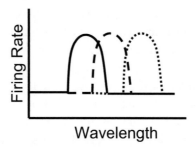

Figure 9.6: Example tuning curve from a cone cells in the eye.

Summary

(1) Neuroscience research is driven forward in large part by the quest to uncover how neurons encode information and how to understand how neurons share information.

(2) Information in the neural system is encoded in the frequency of neural firing.

(3) Extracellular recordings can be used to record signals from many neurons simultaneously. The propagation of an action potential in a neuron creates a series of distinct extracellular deflections.

(4) Analysis techniques can be used to sort the firing of multiple neurons simultaneously based upon characteristic deflections. The firing frequency (information) of each neuron can then be determined. From this data, correlations may be made between neurons.

(5) Tuning curves describe how a population of neurons encodes information in response to a range of stimuli.

CHAPTER 10

Applications

10.1 SCIENCE AND SCIENCE FICTION

A compact definition of science fiction is *science that hasn't happened yet*. Using this definition we can ask when some science fiction of the brain will become science. A quick overview of some of the most popular science fiction books and movies will show that there are (and have been) some incredible leaps of imagination in what people and machines may be expected to do in the future. Although at the time these ideas were far fetched, the pace of breakthroughs is such that some of these wild ideas may not be far from realization.

The purpose of this text is to provide a basic quantitative background from which to jump off to more advance areas of study. Although we cannot go into the details of each of these very interesting research areas, this chapter is an overview of some of the most promising new directions in neuroscience and neuroengineering. As you read each section, try to consider the technology from two perspectives. First, how will the quantitative background of the previous nine chapters inform and supplement these new lines of research? Second, given these new technologies, what ideas do *you* have for future experiments or devices?

10.2 NEURAL IMAGING

The electrical function of neurons was uncovered using primarily patch clamp electrodes and extra-cellular electrodes. These two techniques are very efficient and can detect small changes over a short period of time. Over the past four decades, however, three new imaging techniques have enabled a fresh and complimentary view of the function of the brain.

10.2.1 Optical Recordings

In the late 1960s and early 1970s, the work of Tasaki et al. found that small molecules could be embedded into the cell membrane (Fig. 10.1) that would absorb light at one wavelength (color) and emit light at a different wavelength (left panel of Fig. 10.2). Furthermore, for some molecules, called *fluorescent dyes*, the intensity of the emitted light was proportional to V_m within a range that spanned the normal action potential (right panel of Fig. 10.2). Some subset of these molecules had a time resolution of less than $1msec$, short enough to resolve the action potential upstroke.

A typical imaging system is shown in Fig. 10.3 and consists of two light pathways. In the first pathway (solid line), a laser is used to excite the tissue at a high frequency through a dichromatic mirror. The function of the dichromatic mirror is that it will pass light of a high frequency but reflect light of lower frequencies. The cell then absorbs high frequency photons and emits low-frequency photons. In the second pathway (dotted line), the low-frequency light emitted from the

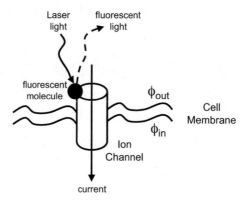

Figure 10.1: Molecular binding of a fluorescent molecule to the cell membrane.

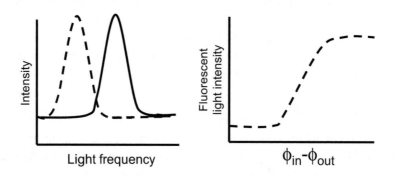

Figure 10.2: Left: Example absorption (dotted) and emitted spectra for a fluorescent molecule. Right: Example of proportionality between V_m and emitted light intensity.

cell is reflected off of the mirror, through a microscope and then to a collection device (usually a CCD camera or photodiode array).

Optical imaging provides at least three advantages over the use of extracellular electrodes. First, as demonstrated in Ch. 8, a single ϕ_e recording is a single weighted average of all surrounding electrical activity. On the other hand, optical mapping enables V_m to be recorded from many locations simultaneously. Second, when coupled with a high power microscope, the field of view can be adjusted to examine populations of neurons, an individual neuron or even regions of a single neuron. Lastly, new dyes have been discovered that are dependent on $[Ca^{2+}]$ instead of V_m, opening up the potential to better understand bursting and synaptic transmission.

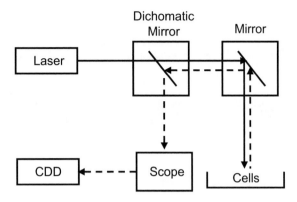

Figure 10.3: Optical mapping hardware.

There are, however, a number of disadvantages of optical mapping. First, the equipment is more expensive than using Ag-AgCl electrodes. Second, the dye itself is somewhat toxic and therefore can not be used for long periods of time in cultures or at all in humans. Third, as light can only penetrate a short distance into tissue, optical imaging is generally limited to thin slices of tissue. Lastly, the response times of fluorescent dyes are orders of magnitude slower than electrodes. Therefore, resolving fast propagation can be difficult.

10.2.2 Functional Magnetic Resonance Imaging

Magnetic Resonance Imaging (MRI) is a technique widely used in neurology clinics for its ability (unlike X-rays) to image soft tissue and (unlike ultrasound) penetrate the bone of the scull. At its most fundamental level, MRI is a three-step process as shown in Fig. 10.4. First a very large (1-3 Tesla) magnetic field, B_0, is applied. The effect is to align all protons (H^+) to the B_O field. Since H^+ ions have a natural spin, they continue to spin like a top even under the influence of the B_0 field. Second, a fast radio frequency pulse, B_1, is introduced perpendicular to the B_0 field. The effect is to knock the aligned spinning molecules into a wobbly orbit. Lastly, the image is created by detecting how these wobbly orbits decay back to being aligned with the B_0 field. The decay rate is called the T_1 time and can be detected by the flow of current induced in a coil of wire. Superimposed upon the T_1 decay is a second decay called T_2. This second decay is a result of interactions between the spin of neighboring molecules. Most importantly in the creation of medical images is that the T_1 and T_2 times are dependent upon the type of material.

There are many subtleties to the MRI technique that have been skipped over, but impressive images of soft tissue can be made quickly in 3D at resolutions approaching $2mm^3$, allowing for fast diagnosis of structural problems. The two most significant disadvantages of MRI are that the technique is expensive and requires very specialized equipment.

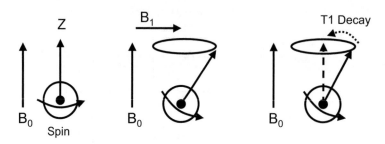

Figure 10.4: Overview of the biophysics of MRI.

Functional MRI (fMRI) is a reinterpretation of traditional MRI that allows for regions with active neurons to be highlighted. Active regions of the brain will demand more O_2, causing oxy-hemoglobin (HbO_2) to become deoxyhemoglobin (Hb). Deoxyhemoglobin is paramagnetic and so has a significant impact on spin interactions and T_2 times. Regional changes in Hb concentration, as a result of neuronal activity, can therefore be detected as a change in the T_2 times. Again, real-time 3D maps can be created. The disadvantage of fMRI is that a large group of neurons must be firing for many seconds before the drop in deoxyhemoglobin can be detected.

10.2.3 Diffusion Tensor Imaging

A second imaging technique that also uses MRI technology is Diffusion Tensor Imaging (DTI). The same physics and equipment are used and the images are similar. But, more information can be added by taking several closely timed images. When the images are averaged together, regions that are static will be reinforced while regions that are in motion will be blurred. Since the brain is not moving, the blurring is caused by the diffusion of water molecules. It has been found that water diffuses much more easily inside the cytoplasm of the axons and less easily across the membrane. In this way, the pathways of axon bundles (tracts) can be identified using DTI as the directions along which the most blurring occurs.

10.3 NEURAL STIMULATION

10.3.1 Cortical Stimuli during Surgery

During neurosurgery, a surgeons will do their best to avoid removing or damaging regions that are vital to the patient's normal cognitive functions. Unfortunately, these areas are not in the same location from one person to the next so the surgeon will often create a *functional map* before beginning the surgery. Creating a functional map is often performed by electrically stimulating a region of the cortex and then asking the patient what they experienced. Remember that neurosurgery is often performed on conscious patients who feel nothing because the brain has no sensory neurons.

10.3.2 Deep Brain Pacing

Recently, the symptoms of Parkinson's Disease and some forms of chronic pain have been reduced by injecting current into specific sites deep in the brain. It has also been suggested that repeated pacing may dampen depression and prevent epileptic seizures. Furthermore, patient controlled stimulation of select locations in the brain stem or spinal cord could lead to better control of basic bodily functions. For example, stimulation of the sacral nerve is an effective means of bladder control.

Building upon the technology of cardiac pacemakers and defibrillators, some companies have already begun to offer deep brain stimulating devices. There are two potential problems with deep brain pacing. First, the blood brain barrier must be compromised to insert the electrodes, potentially opening the way for dangerous bacterial infections and viruses. Second, in some patients an external stimulus can trigger an epileptic seizure.

10.3.3 Magnetic Recordings and Simulation

We have so far only considered *electric* fields. Currents in the brain, however, generate extremely small magnetic fields ($10^{-9} - 10^{-6}$T) that can be recorded on the scalp by a Superconducting Quantum Interference Device (SQUID). An individual SQUID element uses a phenomenon known as the Josephson effect which only occurs when two supercooled superconductors are separated by a thin isolating barrier. The recording of magnetic potentials from the brain is called *magnetoencephalography* (MEG) and is typically performed with an array of SQUIDs.

Just as currents generate a magnetic field, Faraday's law of induction allows a time varying magnetic field to induce current to flow. Therefore, applying a strong time-varying magnetic field to the brain can force current to flow in neurons. Unlike electrical fields which do not easily penetrate the insulating scalp and skull, the magnetic field can extend deep into the brain, attenuated only by distance. The stimulation of the brain through a magnetic field is called *Transcranial Magnetic Stimulation* (TMS).

A pulse or pair of pulses can be delivered that will cause a depolarization and firing of neurons. For example, if the depolarization is in the occipital cortex, a patient will sense flashes of light. Repetitive TMS (rTMS) has been shown to induce longer lasting changes and provides some interesting treatment and research options. Clinically, rTMS may help alleviate some of the symptoms of migraines, stroke, Parkinson's disease, and depression but the mechanisms are unknown. In the research laboratory, rTMS is an effective and noninvasive way of "knocking out" a very specific region of the brain. By focusing a magnetic field of a particular orientation and polarity, it is possible to hyperpolarize a small region of the brain, rendering it unexcitable. If the subject is asked to perform a task that requires that region, they will be incapable of completing the task. Unlike the images created by fMRI, rTMS provides much stronger evidence that a particular region of the brain is actually *used* to perform a specific task.

10.4   DRUG AND GENETIC THERAPIES

Neurons make very heavy use of proteins and amino acids that function as neurotransmitters, enzymes and ion channels. It is therefore not surprising that genetic mutations and changes in protein expression can have an enormous impact on neural conduction. Any breakthroughs in genetic therapies or drug design/delivery will therefore quickly find applications in neuroscience. Two mechanisms will most likely provide the greatest benefit. First, when a mutation has caused a deformation of a protein, an engineered DNA strand could be introduced to the body (typically through a viral vector) with a correction. The virus would then spread the corrected DNA to neurons which would begin producing copies of the corrected protein. Alternatively, a synthetic drug could be designed to perform a desired function. In fact, the action of many current drugs, both legal and illegal, function in a very similar manner. Second, the mechanisms that control protein expression could be altered indirectly through the availability and effectiveness of second messengers. Again, correction of defective second messengers may be accomplished through genetic or pharmaceutical means. In the future, drug and genetic therapies may even be used in combination to correct a defective protein and ensure proper expression.

One very serious issue with delivering drugs and genes to the central nervous system is the *blood brain barrier* (BBB), which severely limits the transport of molecules to the brain. Two technologies may lift this barrier. First, a better understanding of transport across the BBB may allow drug designers to attach tags to the drug that fool transport mechanisms into carrying the drug across the membrane. Alternatively, there may be a way to find small molecules that can diffuse through the BBB that have the desired effect indirectly. Second, it should be possible to build implanted devices (similar to an insulin pump) that can directly deliver drugs to the brain, therefore bypassing the blood brain barrier.

10.5   BRAIN MACHINE INTERFACE

The concept of a Brain Machine Interface (BMI) is an extension of the old idea that a human and a machine can work together to perform a task. When neuroscientists and neuroengineers refer to BMI, however, they mean the direct connection of a human brain to an external machine in a feedback loop. Signals from the brain are recorded either from extracellular electrodes implanted directly into the cortex or from the surface EEG. Although implanting electrodes in the cortex is invasive, it is permanent and allows for many hundreds of simultaneous recordings instead of the 30–50 recordings of the EEG. The signals must then be decoded to parse out relevant information. Many of the techniques of decoding in Ch. 9 are used to sort spikes and generate frequency maps. These data must then be translated into commands to be sent to the external machine. Perhaps the most difficult aspect of BMI is to design a system that performs the translation of the neural code to machine commands. These machine commands, may be the opening of a valve, rotation of a motor or opening of an electrical switch.

The hope of a BMI is that it will be able to adapt to new situations. Similar to neural networks, the BMI must first be trained by sending in inputs to which the desired outputs are known. Using these known input-output pairs, the parameters of the neural-machine translation are adjusted. Next, the model, now tuned, can be sent an arbitrary input and generate an output that is close to desired. In the ideal situation, the person can use their own senses to monitor the progress of the external device. Corrections to the device can then be made by simply thinking (changing neural firing patterns). In a somewhat less ideal situation, the effect may need to be transduced into a signal that the person can evaluate.

BMI was originally envisioned as a therapy to restore motor control to disabled patients, particularly those suffering from ALS, spinal cord injuries, stroke, cerebral palsy, or amputees. The patient would be capable of controlling a device using neural firings that correlate to some thought. It is also possible that the "external" device would in fact be contained within the body. For example, patients with incontinence could have a small valve installed on their urethra that would be directly controlled by thoughts from the brain. The concept of thought controlled devices has many other applications. For example, dangerous jobs (fighter pilot, underwater bridge building), exploration (space missions) or rescue missions could be performed by a machine under direct control from a human brain.

There are a number of hurdles to overcome before BMI can become widely applicable. For complex tasks, such as the delicate finger movements needed to tie a shoe, the amount of information processing and coordination in time is well beyond the present capabilities of a wearable device. Information can only be sent at the speed of light, so exploration of space using BMI will always contain fundamental delays. These delays may also cause some series problems in performing tasks that require fast action such as flying a fighter jet. Despite these obstacles, there are some encouraging finds. In an amazing display of the brain's plasticity, early studies using BMI in animals have shown that after prolonged use, the cortex dedicates some area to the external device. In other words, from the perspective of the brain, the external device has become an additional part of the body.

10.6   GROWING NEURAL TISSUE

Simply stated, the goals of tissue engineering are to: (1) stimulate the natural growth of new tissue inside the body; and (2) artificially grow tissue outside the body. Growing an entire brain is far outside the realm of our current technology. The most significant hurdle is that tissue engineering has not found an effective way to grow three-dimensional tissues. The fundamental limitation is the transport of O_2 and glucose and removal of waste products. In the body, these functions are performed by capillaries which are no more than $100\mu m$ from each neuron. A first step to overcome this limitation is in new materials (e.g., gels) which allow for vasculature to develop along side the neural tissue. To complicate matters, neurons in a 3D structure will most likely require glial cells to perform maintenance functions and possibly participate in electrical impulse communication. Despite these limitations some impressive functions have been reproduced in two dimensional tissues. Below are two examples of the application of tissue engineering in neuroscience.

10.6.1 Nerve Guides

Some of the most debilitating injuries affect only a small population of neurons. For example, quadriplegics and paraplegics often have a gap in their spinal cord that prevents signals from reaching portions of the body below the gap. If this gap could be spanned, all (or most) function would be returned. In the repair of neural gaps, creating *nerve guides* or nerve channels have shown some promise. The idea is to provide a channel that limits the direction that a nerve can grow. In this way two neurons (or nerve bundles) can be forced to connect. Typically, this method will only work to span gaps less than $1cm$. To span larger gaps nerves can be grafted from some other part of the body. A more recent area of research is on the material of the nerve guide. In particular, the nerve guide can be made of a bio-degradable substance which serves its purpose and then is naturally broken down by the body. Some new materials can also be created with embedded growth factors that will release slowly and stimulate both neuron and capillary growth. Lastly, there is evidence to suggest that Hebb's "fire together wire together" idea applies in the peripheral nervous system. Therefore, a continual stimulus down the nerve gap may promote a better electrical connection.

10.6.2 A Biological Flight Simulator

A second application for tissue engineering is to create biological neural networks that will function similarly to artificial neural networks. These networks of real neurons can take in inputs from some external source, process the inputs and send out a meaningful output. Similar to an artificial neural network, the connections between real neurons are plastic, meaning that they can adapt to identify patterns in inputs.

Perhaps one of the most striking and creative examples of the capabilities of a biological network of neurons is the work of Thomas DeMarse. Using circuit boards created from special materials, neurons were grown and interfaced with recording and stimulating electrodes. The neurons, however, were randomly spread on the circuit and allowed to self-assemble. During self-assembly, the network is being trained to fly a flight simulation program in much the same way a neural network would be trained. After the neurons connected, the network could successfully fly the flight simulator without external input. Even more impressive was that given an unexpected input (e.g., a cross-wind, engine failure), the network adapted the outputs to compensate. Growing neurons on a circuit board may have some interesting applications in creating biologically based computers.

10.7 ARTIFICIAL INTELLIGENCE

Artificial Intelligence (AI) is an interdisciplinary field composed of computer scientists, mathematicians, psychologists, philosophers and engineers. There are two primary goals of AI. The first goal is to mimic the function of the brain to do something useful (e.g., drive your car for you). The second goal is to attempt to build an artificial intelligence so we can learn more about the way our own brain works. One fundamental stumbling block to this second goal has been to define what is meant by intelligence. To most, intelligence requires memory, pattern recognition, the ability to generate an orderly sequence of reasoning (logic), the ability to plan ahead and some feedback between these elements. But some have argued that intelligence should include additional elements such as emotions. Still others have argued that intelligence can only arise in the context of a body that can interact with (sense) the world. And perhaps the most perplexing philosophical question of all is does intelligence give rise to consciousness or the other way around?

As the answers to the questions above have been debated for millennia with little progress made, many engineers and computer scientists have approached artificial intelligence from the perspective of utility (the first goal above). Early artificial intelligence research attempted to reduce thinking to the manipulation of symbols by rules. Although this *symbolic* approach produced some interesting applications (e.g., expert systems), it quickly became clear that the number of rules needed to produce anything resembling intelligence was enormous and well beyond the capabilities of a desktop computer. The second approach focuses more on the ideas of connected neurons to build something that resembles the brain in both structure and function. This *connectionist* approach includes neural networks and has made some steps toward matching patterns, storing memories in many distributed locations and the ability to plan ahead. All indications are that as better mathematical models of neuron and brain function are developed, they will find their way into the field of artificial intelligence.

10.8   CONCLUSION

Although it may not be possible to predict which of the above technologies will change the way we understand and use our brain in the future, it is nearly certain that some will influence our lives in profound ways. The road forward will not be easy and as is typical in science not all of the obstacles to progress are known to us at this time. The other certainty is that the brain still has many more secrets waiting to be discovered. With continued study and hard work, you may have a date with scientific history and even add your name to the list of prize winners in the preface.

CHAPTER 11

Suggested Readings

The aim of this text is to quickly place the reader in a position to move on to more advanced topics. These advanced topics may be found in the current neuroscience and engineering literature or in a number of excellent books. Below are books that will supplement and expand upon the concepts presented here. They are by no means the only good references in the field, but rather the texts with which the author is most familiar.

For a more thorough introduction to the anatomy and physiology of the nervous system, *Neuroscience* by Purves et al. is a very good reference. The standard neuroscience book is *Priniciples of Neural Science* by Kandel Schwartz and Jessell. *Theoretical Neuroscience: Computational and Mathematical Modeling of Neural Systems* by Dayan and Abbott is an excellent place to start for a mathematical view of neuroscience. *23 Problems in Systems Neuroscience*, edited by van Hemmen and Sejnowski, is a source of unsolved puzzles in the neuroscience community. For a more detailed mathematical treatment of the electrophysiology in chapters 2-4 and chapter 8, *Bioelectricity* by Plonsey and Barr is a good source. *Biophysics of Computation: Information Processing in Signal Neurons* by Koch is another excellent resource. For a very detailed and clear treatment of ion channels, "Ion Channels of Excitable Membranes" by Hille is a classic. Spiking Neuron Models by Gerstner and Kistler and *Spikes Decisions and Actions* by Wilson are both outstanding introductions to phenomenological models of the neuron. For a more advanced treatment of the computational capabilities of neurons, consult *Dynamical Systems in Neuroscience: The Geometry of Excitability and Bursting* by Izhikevich. The subject of building biophysical networks of neurons is handled nicely in *Neuronal Networks of the Hippocampus* and *Fast Oscillations in Cortical Circuits*, both by Traub. A more qualitative, but also more wide ranging and current view of the dynamics of networks of neurons is contained in *Rhythms of the Brain* by Buzsaki. A series of articles by leaders in the field of neural modeling are contained in *Methods in Neuronal Modeling: From Ions to Networks*, edited by Koch and Segev. The book contains everything from synapse and ion channel models, to branching cables to the building and analysis of large-scale networks of neurons. Artificial neural networks can be more fully explored in *Neural Networks: A Comprehensive Foundation* by Haykin. An overview of the electronics used to record from neural preparations can be found in *Introduction to Electrophysiological Methods and Instrumentation* by Bretschneider and de Weille. Another source of information on the design of biological circuits is *Bioinstrumentation* by Enderle (part of the Synthesis Lectures on Biomedical Engineering). Current topics at the intersection of engineering and neuroscience can be explored in a series of articles compiled in *Neural Engineering*, edited by He. The reader may also be interested in four other texts that are part of the Synthesis Lectures on Biomedical Engineering, *Brain-Machine Interace Engineering* by Sanchez and Enderle, *Neural Interfacing: Forging the Human-Machine Con-*

nection by Coates, *Multimodal Imaging in Neurology: Special Focus on MRI Applications and MEG* by Muller and Kassubek, and *Estimation of Cortical Connectivity in Humans* by Astolfi and Babiloni

Biography

Joseph Tranquillo is an assistant professor of biomedical engineering at Bucknell University where he has been a faculty member since 2005. He received his Doctor of Philosophy degree in biomedical engineering from Duke University (Durham, NC) and Bachelor of Science degree in engineering from Trinity College (Hartford, CT). His teaching interests are in biomedical signals and systems, neural and cardiac electrophysiology, and medical device design. His research interests are in electrophysiology of the heart and brain, high performance biocomputing, and the non-linear dynamics of coupled oscillators. He is an elected member of Sigma Xi and Heart Rhythm, a member of the IEEE Engineering in Medicine and Biology and Biomedical Engineering Society, and serves as a member-at-large on the Biomedical Engineering Division of the American Society of Engineering Education. When not teaching or doing research, he enjoys improvisational dance and music, running marathons, backpacking, brewing Belgian beers and raising his two children Laura and Paul.

Index

Printed in the United States
by Baker & Taylor Publisher Services